THE COMPLETE SOFTWARE PROJECT MANAGER

MASTERING TECHNOLOGY FROM PLANNING TO LAUNCH AND BEYOND

Anna P. Murray

WILEY

Published by John Wiley & Sons, Inc., Hoboken, New Jersey.

Published simultaneously in Canada.

For general information on our other products and services or for technical support, please contact our Customer Care Department within the United States at (800) 762-2974, outside the United States at (317) 572-3993 or fax (317) 572-4002.

Wiley publishes in a variety of print and electronic formats and by print-on-demand. Some material included with standard print versions of this book may not be included in e-books or in print-on-demand. If this book refers to media such as a CD or DVD that is not included in the version you purchased, you may download this material at http://booksupport.wiley.com. For more information about Wiley products, visit www.wiley.com.

Library of Congress Cataloging-in-Publication Data

Names: Murray, Anna, 1966- author.
Title: The complete software project manager : mastering technology from planning to launch and beyond / Anna P. Murray.
Description: Hoboken, New Jersey : John Wiley & Sons, Inc., [2016] | Includes index.
Identifiers: LCCN 2015036771 (print) | LCCN 2015040806 (ebook) | ISBN 9781119161837 (cloth) | ISBN 9781119219910 (ePDF) | ISBN 9781119219903 (ePub)
Subjects: LCSH: Computer software—Development. | Software engineering–Management.
Classification: LCC QA76.76.D47 M877 2016 (print) | LCC QA76.76.D47 (ebook) | DDC 005.1–dc23
LC record available at http://lccn.loc.gov/2015036771

Cover Design: Wiley
Cover Image: © shuoshu / Getty Images

Printed in the United States of America

10 9 8 7 6 5 4 3 2 1

To my partner in life and business, Chris Moschovitis.
Your support makes all things possible.

CONTENTS

FOREWORD

A generation ago, Internet pioneer Carl Malamud took issue with people who complained about bulky computers, byzantine interfaces, and buggy software: "A lot of this 'too hard to use' stuff will go away. Radio was so messy for the first 20 years, it wasn't funny. Cars, ditto—you had to be a mechanic to drive one." Computers, he prophesied, would one day be as easy to use and reliable as automobiles had become, and the electronic frontier would become as simple to navigate as an Interstate with a road atlas.

That day came, of course—including Google Maps to replace the road atlas—but it brought along a paradox. The simplicity, power, and awesomeness that users and executives take for granted disguise the very technical, messy, and difficult work that happens behind the scenes. Gremlins lurk under the lid of your laptop; krakens crouch behind the scrim of your Cloud. A lot of what seems to gleam is the shimmer of WD-40 on duct tape.

Amateurs cheerfully rely on smoothly running software for business and pleasure, but the work of IT is no place for an amateur. Creating and designing systems, maintaining them, and running projects to install, fix, rip out and replace, upgrade, and integrate them—these are complicated tasks. They rarely go according to plan, because they invariably run into some obstacle that could not possibly be foreseen. Even professionals in the industry can be seduced by a project's progress into thinking, "No problem," when they should be thinking, "Oh, this is hairy." And, of course, finger-pointing is the consequence of problems, with IT blaming the business, the business blaming IT, and everybody blaming the consultants and software companies.

Anna Murray's book is an extraordinary accomplishment: It speaks as clearly to the IT professional as it does to executive civilians like me. (I'm an English major with a Mac.) For the professional, it is an experienced, no-nonsense guide: the mentor you wish you had. Anna tells you how to put an IT project into the strategic framework that is right for your organization, how to scope a project, how to communicate and plan with stakeholders; she tells you how to assemble your tools and team; she explains how to find, select, and manage vendors; and she tells you how to cook and eat crow when you need to—as you almost certainly will.

The strategic point is particularly valuable. IT long ago migrated from the glass house to the desktop, but it is now part of every nerve and sinew of an organization: It is in the skin that touches customers, in the brain that analyzes performance, in the heart that pumps resources where they are needed. Technologists don't need to be strategists, but they can no longer be strategically ignorant. Anna shows how to connect strategy and IT, how to ask the right questions, and how to frame the trade-offs and decisions that complex IT projects require in a way that business leaders can understand.

This is, at the same time, the book IT managers should give to their nontechnologist colleagues, bosses, and internal customers. Her distinction among IT projects that are simple, complicated, and complex is worth the price of admission by itself. I can think back to half a dozen cases where the business-IT relationship got into trouble because the nontechnologists did not know or could not understand what their IT colleagues were up against. Any executive who reads this book will ask better questions, get better answers, and have a better understanding of the answers.

Thomas A. Stewart
Executive Director, National Center for the Middle Market
Fisher College of Business
The Ohio State University

ACKNOWLEDGMENTS

This book grew out of a two-decades-long conversation with my colleagues and with the many clients we are privileged to serve. Starting back in the Windows 3.1 days, we have grappled with the complexities of technology and survived the numerous inevitable crises that accompany software development. My work has provided me with incredible opportunities to collaborate with and learn from these exceptional professionals.

I am grateful for Denise Mitchell and Rebecca Harrigan, incomparable technology leaders, partners, and clients, and for the great team at Kellogg Corporation. Also, for the executive group and the great IT, BA, and PM teams at Harvard Business Publishing. I also owe a debt of gratitude to my own team, from whom I am always learning: Frank G. Andrews, Pedro Garrett, Atsushi Tatsuoka, and Steve Vance.

Thanks goes as well to fellow technology writer Mike Barlow, who showed such generosity of spirit and time in championing this book. To Sheila and Gerry Levine, whose wise counsel has supported me through everything from content to contracts. And, to the International Women's Writing Guild, whose sisterhood of writers and teachers filled all kinds of gaps from craft to the business of writing.

What could be more valuable to a writer than an editorial partner with skill, wit, and wisdom? Hilary Poole, writer and editor extraordinaire, was one of the original collaborators on this project. Thank you for all your help from tidied-up commas to slashed technology jargon.

To the incredible team at Wiley: Sheck Cho, Pete Gaughan, Connor O'Brien, Michael Henton, and Caroline Maria Vincent. Your professionalism and collaboration are all any author could hope for.

And finally, to Chris Moschovitis, whose brilliance in technology and so many other things awes me every day. I am lucky to go to work with you each morning and to come home to you each night.

ABOUT THE AUTHOR

Anna P. Murray is a nationally recognized technology consultant and the founder of emedia LLC, a certified women-owned business. She has consulted on and run large-scale software projects for Kellogg, Harvard Business Publishing, Time Out New York, Slate Group, The American Association of Advertising Agencies, National Cancer Institutes, *New York* magazine, and many others.

Murray is a double winner of the Stevie Award for Women in Business, a recipient of a Mobile Marketing Association award for mobile app development, several Kellogg agency partner awards, and Folio's Top Women in Media Award. She has served as President and Vice President of the International Women's Writing Guild.

One of a rare species—a woman who owns a successful software-development company—she loves to combine her two passions: technology and writing. Anna writes on technology from a variety of vantage points, including its humorous impact on daily life, its serious business applications, and its role in changing the way people relate to one another.

After spending several years as a teacher and journalist, Murray founded emedia, one of the earliest web-development firms, in 1996. She holds a bachelor's degree from Yale and a master's in journalism from Columbia.

INTRODUCTION

The history of software development teaches us that between 30 and 40 percent of all software projects fail. A good majority of those are canceled completely and never see the light of day. It seems everyone has a different prescription for how to avoid this fate. There are shelves of books on how to improve the software development process—books on Agile methodology, Scrum, the Waterfall method, rapid application development, extreme programming, top-down versus bottom-up, and even something called the chaos model.

When a business is faced with a software development project, people don't have the time to become experts in software management and development theory. What you need is not theory, but rather a practical, hands-on guide to managing this complex process, written in a language you can understand.

Some undertakings—be it having kids, climbing Mount Everest, or, yes, software development—are just difficult. And sometimes the most helpful thing is to hear from someone who has been there before and walked the path. That person can tell you what you're going to encounter and what difficulties to expect. An experienced guide may not be able to take away the challenges, but she can prepare you for them.

About a year ago, I was working on a large implementation project. About halfway through, a young project manager came to me and said, "Anna, do you have a crystal ball? Because at the beginning of this project you sat us all down and told us what to expect. You gave us a list of many things and every single one of them has happened—and happened on schedule. Do you have that written down anywhere?"

Now I have it written down somewhere.

Software Project Management

This book is specifically about software project management. There are many other types of project management. Project management is needed for digging oil wells, building skyscrapers, and launching rockets.

Examples of software projects include

- Launching websites
- Installing a CRM (customer relationship management) tool
- Implementing a new accounting package
- Building a custom application for your business

Software is truly "its own animal." People tasked with managing software projects need processes and advice just for them.

A Holistic Approach

Many books on project management focus on what I call the "literal project manager," the person whose title is "project manager" on his business card. He or she may have a specific certification in the discipline of project management. In Chapter 5, we'll talk in depth about project teams and team members. For now, it's important to understand that the literal project manager is not the only one with project-management responsibilities. People might even be surprised learn how narrowly focused the project manager's job often is (Figure I.1).

Many books on project management concentrate only on the literal project manager. The purpose of many of these books is to help the project manager pass his or her certification tests, such as the PMP exam. That's great, if you are facing the exam. Unfortunately, this narrow focus leaves much out, not only for the literal project manager but for everyone else.

True project management extends to many people and to the entire business. There is the project sponsor, the program manager, the project manager, and programming teams as well as stakeholders in the broader business. Therefore, this book takes a holistic view of project management. It can be read by businesspeople, programmers, project managers, program managers, and CEOs—anyone who is involved in or affected by a software project (Figure I.2). Of course, it has critical information for people whose job titles are project manager and program manager.

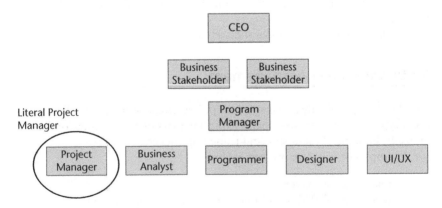

Figure I.1

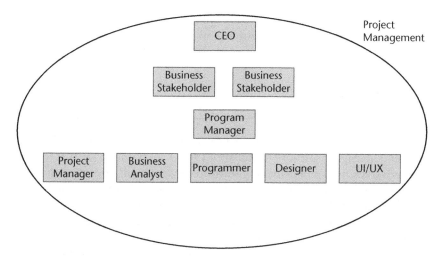

Figure I.2

For Medium-to-Large Projects

Project management looks different depending on the size of the project. The sort of project management involved in putting up your personal WordPress blog likely happens all in your own head. Inside of businesses most projects that last more than a couple of weeks require some kind of external project management.

For this book, I will be giving processes and advice appropriate for medium-to-large projects. This generally means projects that take at least three months and may last as long as two years. The reason I've chosen to focus on projects of this size is they are the most common. Also, smaller projects, as noted earlier, probably don't need much project management at all. Finally, I have found that once people understand the fundamentals of managing medium-to-large size projects, they are able to scale the processes and recommendations both up and down to suit their needs.

Agile vs. Waterfall

For those of you who don't know these terms, don't worry. We'll talk more about this in Chapter 2. Many people familiar with modern software development want to know right up front: Are you talking about project management practices for Agile development or for Waterfall?

This book assumes the incorporation of key Agile methodologies, while acknowledging that most businesses cannot adhere to a purely Agile style. In fact, the issue of Agile vs. Waterfall is a major reason I wrote this book. After running hundreds of software projects inside many different types of businesses, I discovered that good project management is a blend. It must be "agile enough."

Why Listen to Me?

As the CEO of emedia (www.emediaweb.com), I have been developing software and managing software teams since 1994. I am proud to say that we have a superb track record in delivering software development projects, from large database implementations to small business websites, on time and on budget.

I've managed website launches, database migrations, finance system replacements, custom application development, content management system implementations, and packaged software deployments. The dollar value of these projects has ranged from the small thousands to the tens of millions. It's allowed me to see the commonalities among many software development endeavors and to develop strategies to improve the process.

I am also proud to say that the majority of my customers have been with me for years—a decade or longer in some cases. When you have a challenging undertaking, you want a partner you trust.

Who Is This Book For?

If you're inside an organization undertaking a software development project, this book is for you. If you are a consultant or vendor who rolls out software for customers, this book is for you, too. If you are a business leader, programmer, project manager, or program manager, this book is also for you.

This book has chapters on all the pragmatic stuff you need to know: organizing and staffing a project; the common conflicts that crop up; planning; risk management; and, of course, budgeting. There's even a chapter for people who are midway through a project and are encountering problems. If you read from start to finish, the book will take you through the life cycle of a software development project. Or, you can dip into the chapter based on your particular interest or area of challenge.

CHAPTER 1

Software Development Explained: Creativity Meets Complexity

There are shelves of books and hundreds of thousands of articles dedicated to making software development better. Why has it been so hard for smart professionals to just make software projects run smoothly, on time and on budget? What's up here?

A Definition of Software Development

Software development is any activity that involves the creation or customization of software. As noted in the introduction, it can include

- Launching websites
- Installing a CRM (customer relationship management) tool
- Implementing a new accounting package
- Building a custom application for your business

All these activities qualify as software development. Most businesses will, at some point, be confronted by a software development project. Technology is now so intrinsically integrated into business that it's impossible to avoid.

Why Is Software Development So Difficult? Hint: It's *Not* Like Building a House

A lot of people use the metaphor of house building as a comparison for the activity of software development. I believe this metaphor does an enormous disservice to the process. I reject this metaphor because it gives people a false sense of security and a false understanding of the nature of software development.

A house is concrete and well understood by all. We have all been inside houses. We all share comparable assumptions about what a house is. The same cannot be said for software. In many cases, I have sat in a room with people with completely divergent views about even the most basic aspects of their software project.

Frequently, in developing software, you are creating something from nothing. That means the end product could be practically anything. Here are some endeavors in which, as in software development, you are creating something and the outcome could be a wide range of possibilities:

- Writing a novel
- Growing a garden
- Composing a symphony

There is a "blank slate" quality to creative activities. When you begin a novel, for example, the end product could take an almost infinite variety of forms. The listed endeavors, you'll notice, are all open to a wide scope of interpretation. One person's assumptions regarding the nature of a garden (flowers) may not remotely match another person's (vegetables). So it is with software development: You might think that the parameters of your project ought to be "obvious"—but they may not be obvious to your colleagues.

The previous examples all capture a fundamental reality of software development. By its very nature, software development is creation. You're going from a state where something doesn't exist to one where it does. At the beginning, the outcome could be anything, which means that everyone in the room probably has a different understanding about what the project actually is.

The Simple, the Complicated, and the Complex

In *The Checklist Manifesto* (2009, Metropolitan Books), a book I highly recommend, author Atul Gawande talks about three types of endeavors—the simple, the complicated, and the complex. It's helpful to understand these distinctions because software development almost always involves all three:

- Simple project components are easy to conceptualize. You know what needs to get done, and you simply need to get out the elbow grease and do it.
- Complicated project components are hard to understand and involve a lot of steps, but they are not very risky. If you read the directions carefully enough and follow them, you will get the project done.

▣ Complex pieces of projects, on the other hand, have a lot of variables like the complicated, but they are also highly fluid and very risky.

As already noted, software projects almost always involve all three types of activities: the simple, the complicated, and the complex. The percent mixture of each depends on the project.

I use three of my own metaphors to describe the simple, complicated, and complex as they relate to software development. Mastering these three metaphors and learning to apply them in software will help you manage any software project more successfully.

Metaphor #1: Piles of Snow

"Piles of snow" is the phrase I use to describe the *simple* activities within a software project.

Imagine a massive snowstorm blew through overnight. You wake up and your 200-foot-long driveway is completely blanketed in white. Worse, the plow guy called to say he can't make it. The city plow comes through and there is now an even greater pile at the end of your driveway (Figure 1.1).

What you need to do is absolutely clear. Get a shovel and dig! How to do it is also no mystery.

Keep in mind *simple* doesn't mean *easy*. In fact, most of the time a lot of labor is involved in piles-of-snow software tasks. It's going to be a lot of work

Figure 1.1

to shovel that driveway, especially if there's no one to help. But the "project" is simple in concept and in execution: Dig until you are done.

I will return to the "piles of snow" metaphor again and again in this book.

Metaphor #2: The Ikea Desk

"Ikea desk" is a term I use to describe *complicated* aspects of software projects.

Furniture from the Swedish store Ikea comes boxed up and involves seemingly millions of little pieces (Figure 1.2). The directions are expressed solely in pictures, presumably because it's more efficient to do it this way when you serve an international customer base. Imagine the effort to translate all those directions into hundreds of languages!

To begin your Ikea desk project, you must lay out all the tiny pieces on the floor and match them to the illustrations in the directions. Then you must meticulously follow the directions. There is frequently trial and error. Wait. Was that the part? It doesn't seem to fit. No, I think this one is the right screw. It's the smaller-than-middle-sized one.

Anyone who's put together an Ikea desk remembers a weekend spent on the living room floor before the furniture piece is finally ready. It can be frustrating. It's time consuming. You may be missing a part and have to go back to the Ikea store and wait on the customer service line. Despite all this, success is virtually guaranteed. With enough time and Allen wrenches, you will get it done.

Figure 1.2

Ikea desks require more application of brainpower (e.g., reading and interpreting detailed directions), more concentration, and more backtracking and redoing than shoveling piles of snow. Further, the time it will take to assemble the Ikea desk may be harder to judge than the driveway shoveling. You may say something like, "I'll have this baby assembled by noon," only to realize you put the wrong screws in and assembled it backwards. It's more like midnight when you actually finish. But in both cases, a successful outcome is 99 percent guaranteed.

Metaphor #3: Heart Surgery

The Checklist Manifesto uses heart surgery as an example of a *complex* activity, and it works well to describe aspects of software development.

To perform heart surgery, you absolutely need extensive training and skill. Further, your patient might have an undiagnosed medical problem that causes the operation to unexpectedly fail. The human body is not 100 percent understood by anyone. It is squishy and organic and unpredictable. The fact is, no matter how much you plan, certain variables are out of your control (Figure 1.3).

Unlike Ikea desks or piles of snow, heart surgeries are unpredictable and risky. Training and experience helps, but even with all that, you have no guarantee that the surgery will succeed.

Hint: Finding a surgeon with experience in risky heart surgeries does help.

Figure 1.3

Using the Three Metaphors in Project Management

Ideally, your project would involve many piles of snow, a moderate number of Ikea desks, and as few heart surgeries as possible. You would strive to reduce complexity in your software to the degree possible. In many cases, this is the best approach and we'll continually return to it throughout this book.

But sometimes the choice isn't up to you. Your project may simply have complex elements you can't avoid. Many, if not most, software projects do.

Furthermore, to say something is "complex" or a "heart surgery" is just another way of saying it's risky, with many unpredictable and unknown elements. This brings us back to the concept of creativity, discussed at the beginning of this chapter. Creative endeavors involve, by definition, lots of unknowns. One of the main reasons people undertake software projects is, indeed, to be creative and to come up with something new. That's what we call innovation! And innovation is a key business goal for many organizations. Is there a business that does *not* wish to pursue innovation?

The problem comes in when people sign up for complexity in an unconscious way. They never look at the pieces of their project and evaluate which are snow piles, Ikea desks, and heart surgeries. Worse, believe it or not, some people seek out complexity. Why would someone do that? Signing up for complexity and risk sounds crazy, right?

One only needs to drive on the freeway for a reminder that human beings often and consistently seek out risk. Software development is usually exciting to business stakeholders, who may spend most of their lives in dusty old Excel spreadsheets. The chance to do something creative and techy brings out the risk taker in many people.

The reality is that most people, even programmers, do not analyze the projects they undertake in terms of the simple, the complicated, and the complex. They sign up for a project and simply accept, "It is what it is." However, especially in the first phases of your project, it is essential to identify what is a pile of snow, what is an Ikea desk, and what is likely to be a heart surgery. Then you can make informed decisions on project approach, planning, staffing, and budgeting. You can make considered choices about where you can and cannot reduce complexity and where you consciously wish to sign up for complexity to achieve your business goals.

CHAPTER 2

Agile, Waterfall, and the Key to Modern Project Management

Agile and Waterfall

Many professionals, especially those with little or no technology background, enter into projects in an unconscious way. They may not have heard the terms "Agile" or "Waterfall." And they might not know which of these methods their tech teams practice. Agile vs. Waterfall is certainly the most commonly heard debate in terms of approaching a software project.

People involved in software development have almost certainly heard the terms Agile and Waterfall. They will be familiar with the ongoing debate about these two forms of project approaches. Understanding these two ways of approaching projects—when and where one method has an advantage over another, and what version in what balance is right for your organization—is foundational to good software project management.

First, here are some definitions.

Waterfall

You could replace the term "waterfall" with the term "traditional." The Waterfall project approach is generally regarded as a more traditional way to manage software development.

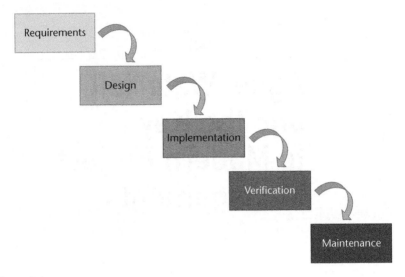

Figure 2.1

The Waterfall process has been around for decades. It is a sequential way of developing software:

- Gather software requirements
- Design the technical solution
- Code the software
- Test and debug the software
- Deliver the finished product
- Enter maintenance mode

Sounds logical, right?

The name Waterfall came from the boxes and arrows used to illustrate these steps. The arrows make it seem like water is flowing over a series of small waterfalls (Figure 2.1).

Waterfall's Problems

Nothing is perfect and the Waterfall method is no exception. Over many decades Waterfall was the default method to develop software, and many problems were common.

The Requirements Requirement

The Waterfall method calls for all the software's requirements to be thoroughly collected and documented up front, then handed off to an implementation team. This practice conjures an image: Tome-like requirements documentation is presented to a programmer who then disappears into a cave for long months before emerging in spring with a finished product.

It's very difficult to do "thorough enough" requirements gathering. How much is the right amount? You could go on infinitely documenting every possible user click under every possible condition.

Two results were common in the Waterfall method: (1) Requirements took so long as to be impractical. (2) Requirements were missed. In the first case, a software project can't get off the launch pad. In the second, programmers may work while assuming they have the complete requirements, only to discover later they were wrong. Since the product is not seen until the very end in the Waterfall method, this leads to costly refactoring of software.

Inflexibility

In the Waterfall method, you define things first, and then the programmers get to work. Changes to specifications are frowned upon. There is a formal "change request" process by which any alteration to the initial requirements goes through a review and costing process.

Loss of Opportunity and Time to Market

One of the famous rules in technology is Moore's Law. This is the axiom that states that computing power will double every 18 months. You see the effect of Moore's Law all around you. Chips get smaller, cell phones get more powerful, and flat screen TVs get cheaper.

Moore's Law also has an impact on software development project management. In technology, things move fast—really fast. If a software project takes 18 months to deliver, you've missed an entire Moore's Law cycle, because the business specifications were written 18 months ago.

What has happened in the intervening months? What progress has been made in the technology world around us that your project could leverage?

If you ask these questions in a formal Waterfall environment, you'll be greeted with a snarl and presented with a change-request form.

Customer Dissatisfaction

Did the project requirements truly capture what the business wanted? It takes a very talented and thorough business analyst to do this. Were the requirements accurately transmitted to the programming team? Again, very nuanced and consistent communication must happen to achieve this goal.

Here's another reality: People forget. It's unfortunate, but they do. For a little while, as the project team interacts with the business stakeholders and gathers requirements, the business stakeholders are deeply engaged. Then, the project team disappears to work on the project and the business stakeholders forget. Worse, they often start to embellish in their own heads what the software will do when it's done. The finished business requirements document is often never read. The business stakeholders look at the impressive pile of paper and just assume it contains all their hopes and dreams. It must, right? It's so thick with so many Visio diagrams. The business requirements document collects dust in a drawer.

When the project is delivered, many months later, the customer declares it wasn't what they wanted.

Agile

"Agile" is a relatively new term, officially coined in 2001 when a group of software professionals released a document called the *Agile Manifesto*. However, many of us who were in software development prior to 2001 were already working on ways to make software development more flexible and reduce some of its problems.

The *Agile Manifesto* (and many books and articles written since) codifies a software development method that isn't just a modification of the Waterfall method. It throws Waterfall over the proverbial waterfall.

The easiest way to understand Agile is as the extreme opposite of traditional sequential software development. The Agile process values constant collaboration, frequent deliverables, and continuous evolution of requirements. There is constant communication; shifts are made in real time; and surprises are less frequent.

Figure 2.2 shows once again a diagram of the Waterfall process. Contrasted with this is the Agile diagram shown in Figure 2.3.

In Agile there is lots of circling back, doing, and redoing with customer input to get it right.

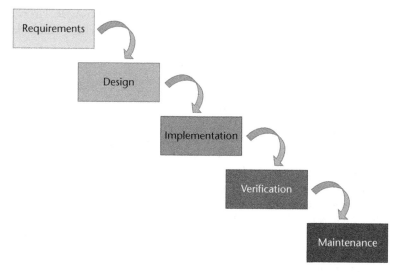

Figure 2.2

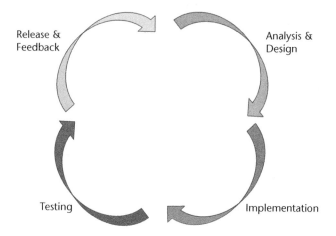

Figure 2.3

Agile projects

- Provide frequent deliverables to the end customer for review
- Value changes and input through the life cycle of the process
- Gather requirements on the fly
- Have lots of collaboration, including a daily meeting with all of the team members

There's no disappearing into a programming cave with Agile. Everybody is in constant collaboration about the end deliverable and change and adaptation are the names of the game.

Not surprisingly, Agile isn't perfect either.

Lack of Up-Front Planning

An Agile project starts with only the most high-level requirements. Sometimes these are referred to as "user stories." Such a requirement might sound like, "A user will be able to buy a subscription to our product on a new e-commerce website." There are no designs, no specifications.

The project team begins work immediately, quickly whipping something together such as the first page the user will see with a list of the possible subscription levels. The team presents it to the business stakeholder who gives feedback. You might think of Agile as a process of "You're getting warmer. You're getting warmer. Oops, now, you're getting colder. Oh! Now you're red hot."

This process can and does produce great and innovative software. However, inside a business, allowing for such flexibility can be hard, and many times it is impossible.

For example, your leadership or board of directors might want to know in advance what you plan to do and how long it's going to take. In my experience, most insist on it. In the Agile method, you can only give the most general answers.

Lack of Up-Front Costs

If you can't say with specificity what you're building or how long the project will last, you also can't say how much it's going to cost. Agile proponents will say, correctly, that practitioners of traditional software development methods (Waterfall) get the budget wrong more often than not. That's true, and this book has recommendations on how to deal with budgeting in Chapter 8.

Agile's solution to the inaccuracy of software budgeting is to avoid the topic almost completely. A common scenario in Agile is that a team is established, and their monthly expenses are accounted for across some period of time that someone guesses will be the duration of the project. Then the team begins working and works until they are done.

Needless to say, this method of costing projects does not fly within most traditional American businesses. If an important software initiative has been identified, the first thing the CEO and board want to see are the plans defining what the software is, the breakdown of specifications, and a detailed budget of costs.

The lack of budgeting alone will disqualify Agile for many organizations. How can an executive sign off on a project for which there is no precise budget and scant documentation? In the Agile method, an enterprise must commit to a very loose budget and tolerate minimal specifications about what will be achieved and when.

Stakeholder Involvement

Businesspeople throughout the organization have a stake in the software development process. Sometimes they are the end customer, as in the case of the delivery of a new accounting system or CRM system. Sometimes they are a proxy for the end customer, as when the sales and marketing department requests a new e-commerce website that allows people to buy the company's products more easily. In either case, these businesspeople are the "key stakeholders" or "internal customers" for the project. We'll talk more about teams and roles in Chapter 5.

Both methods, Waterfall and Agile, require the attention of business stakeholders. Of course! They are the ones for whom the product is being delivered. But, the Agile method requires a lot *more* attention, consistently, sometimes over the months or years it takes to deliver the product. In Agile, the business stakeholder attends the daily Scrum meetings and gives feedback on all the micro-deliverables prepared by the programming team.

What about that new customer you must visit on a four-day trip to Dallas? Or that board of directors meeting in Miami? Or that two-week trip to India to investigate the Asia Pacific market? How is the programming team to move forward in your absence? The Agile method often breaks down under these real-world conditions.

Extensive Training

It *is* actually possible to ride a horse without a saddle or a bridle in an open field. But you have to be a pretty good rider to do it. Freedom from restraints such as corrals and riding gear takes years of training and practice. Try it as an amateur, and you'll likely get hurt. Similarly, there is almost nothing riskier than badly implemented Agile. Project teams meander around without direction from unavailable business stakeholders who don't understand their role in any case. Dollars evaporate and goals are never reached.

Do a quick Amazon search on Agile and you will find scores of books on how to train for Agile methodologies inside a business. It can take years to train to be a "Scrum master." I know companies so determined to become more Agile, they hire consultants to instruct their tech teams and business stakeholders for months so that Agile can be implemented correctly.

Where Agile Works Best

Agile works very well inside of tech companies where the end product is a piece of software, in contrast to companies where tech plays a supporting role. In companies such as these, speed and innovation are at a premium. Both company leaders and team members may already have experience with Agile, eliminating the learning curve. Silicon Valley CEOs and boards are accepting of the lack of definition, risks, and imperatives to "just get going." Such a company has programmers already on staff. In this case, there is no reason to sit and wait while business analysts collect requirements. Getting the programmers to deliver a "beta" product and test it in the marketplace makes all the sense in the world. Gmail is a great example.

Another place where Agile is a good fit is after a large software product has been delivered and you are in support/maintain/enhance mode. In this case, the business has baked in some costs for ongoing maintenance and support, so the Agile budgeting problem is largely solved. Also, the Agile practice of quickly prioritizing, estimating, and tackling a list of backlog items works well. We'll cover support and maintenance more in Chapter 13.

In general, Agile is better suited to projects that

- Are smaller, less complex
- Have constant availability of stakeholders
- Have few complex integrations needing advance planning
- Have leadership accepting of cost and scope changes
- Need very fast time to market

The Need for Up-Front Requirements in Many Projects

Laying out business requirements is usually an absolute necessity, especially when your project has certain high-risk characteristics. Take integrations.

An "integration point" is where one system exchanges data with anther system. Let's say, for example, your company has a membership database and you want your members to log in and receive special privileges if they are members. In this case, the website will have to call over to the membership system and ask, "Hey, membership system, is this person a member? Can they see the content they're asking for?" If the membership system gives a yes answer, the website serves up the appropriate content. If the system says no, the user may not proceed. This is an integration example. Integrations are fundamental to most software development.

What if you start to develop your website, in the previous example, using a tool that integrates poorly with your membership system? What if the system you are trying to integrate with has a certain kind of technology your team is unfamiliar with? What if a different approach to the website would have made this all easier? Without the gathering of up-front requirements, you wouldn't know any of this.

There are many reasons to do up-front research and planning. In addition to its importance for budgeting, it prevents surprises later. Up-front planning also leads to fewer coding flaws, more robust software architecture, and fewer security holes.

The Real World

The fact is that Agile is still quite uncommon inside of most organizations. Many organizations say they "use Agile," but many times they are only incorporating a few details like the daily stand-up meeting Agile is famous for.

The methods discussed in this book weave in the most realistic and achievable Agile principles, the ones that fit into the more traditional software development situations inside the companies where most people are likely to work.

Agile Enough

As noted earlier, many people will attest to Agile's benefits, but this fluid method is rarely achievable inside most organizations. In this book, I will give you a practical path to greater success managing software projects in the real world.

Recent research suggests that a blend of the two methods is the best way to go in terms of both cost and the quality of the end deliverable. In particular, the Waterfall emphasis on up-front design ensures a product with stronger code and fewer performance problems and security weaknesses.

The Software Development Life Cycle

Like all creative activities, software development has a life of its own. The term has been in use since the 1960s, and it is graphically represented much

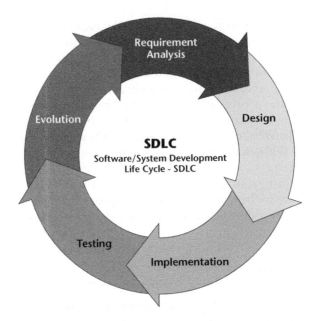

Figure 2.4
Source: © 2012 Cliffydcw. Used by permission of Creative Commons license
(https://creativecommons.org/licenses/by-sa/3.0/).

like the circle of life (Figure 2.4). It recognizes the following phases: Analysis, Design, Implementation, Testing, and Evolution. Of course, there are many other variants (e.g., replacing "Analysis" with "Requirements" and "Evolution" with "Maintenance") but the point remains the same: Software has a life cycle, and that life cycle is intractably woven with the life cycle of your business.

CHAPTER 3

Project Approaches; Off-the-Shelf and Custom Development; One Comprehensive Tool and Specialized Tools; Phased Launches and Pilots

The phrase "project approach" covers several topics that usually need to be considered at the very start of a project. Some subjects under project approach touch on technological philosophy. You may not have a philosophy yet. Don't worry! We're about to introduce you to some key viewpoints. Another aspect of project approach deals with the high-level technology decisions that must be made at the beginning of projects.

As you embark on your project, it's important to evaluate and decide upon what *your* approach will be. Deciding project approach means addressing the following questions:

- Will I use an off-the-shelf product, or custom coding?
- Will I choose a product that does many things, or a specialized product for each aspect of my software project?
- Will I launch the software all at once, or in phases?
- Will I pilot the software first before launching?

These questions are critical to project planning and execution. The questions may not be answerable up front. Your research and planning phase, covered in Chapter 5, is often critical to finding the questions' answers. But it's important to be aware of the key project approach options before you start.

The Custom vs. Off-the-Shelf Approach

Most software projects start with planners and engineers asking whether there is some off-the-shelf piece of software that addresses the business requirements, or whether the job will need "custom coding." The custom vs. off-the-shelf topic is a debate you will often hear at the start of projects. It's rare to begin a project without this project approach consideration surfacing.

History

Through the 1980s and 1990s, if you wanted a piece of software developed, it pretty much had to be custom coded. That means a programmer would be working "from scratch" to write all the lines of code to make the thing work. This process was both laborious and expensive. Companies with custom software needed programmers consistently engaged, either on-staff or in long-term consulting relationships, to manage, update, and troubleshoot the custom software that was now a key part of business operations.

In the late 1990s and through the first decade of the 2000s, more and more off-the-shelf software became available. A good example of off-the-shelf software is Salesforce.com, the customer relationship management (CRM) platform.

Why would you build your own software if you could just buy Salesforce? Further, with an off-the-shelf piece of software, companies don't have to employ their own programmers. Salesforce does all that! It's Salesforce's problem to update, configure, improve, and patch the software. Good off-the-shelf software emerged for nearly every business function and across every industry. Prices dropped, businesses benefited, and it seemed crazy to engage in custom development anymore except for the most innovative, specialized, or unique projects.

Off-the-shelf software seems to guarantee you a project that is made up of only "piles of snow" and "Ikea desks."

The Benefit of Off-the-Shelf

The promise of off-the-shelf software is that it's pre-built. You tick a couple of settings that match your business needs and off you go! QuickBooks, which many people are familiar with, is an off-the-shelf product example. Obviously you must put in your own information, but the software gives you all the tools you need to do accounting for your business.

Remember the discussion about the creative nature of software in Chapter 1? You could say with off-the-shelf software you are "editing" something that exists rather than "creating" something from scratch.

The trade-off with packaged software, as you might expect, has to do with risk and reward. Configuring something that exists is inherently less risky than creating something new. Right now, you may not fully understand what "risky" actually means; by the end of this book, you will. For now, let's just say risky means having to explain to your boss, board, or investors that you are going to be late delivering the project and it is 40 percent (or more) over budget.

Off-the-Shelf Examples

The QuickBooks example, as with many accounting packages, is a great illustration of a piece of off-the-shelf software. Almost no one develops accounting-type software from scratch. Running one's finances is a fairly well-prescribed activity. Generally Accepted Accounting Practices (GAAP), the IRS, banks, and other financial entities have helped to create a world where the business requirements are extremely clear. Therefore, QuickBooks is also a good example of off-the-shelf software requiring a limited amount of change or customization. Put another way, everyone means pretty much the same thing when she asks, "Can you run me the P&L for last year?"

Occasionally you may encounter an office that is not even computerized. This office may be accustomed to a quirky way of doing things. But in cases like these, when the decision is made to adopt an accounting package such as QuickBooks, the company shifts its practices toward what is supported in the off-the-shelf product. They rarely say, "We should build our own accounting package to match the way Mary records things in her ledger."

E-commerce is another good example of off-the-shelf software that works in a very defined way. Off-the-shelf e-commerce systems come with an "out-of-the-box" shopping cart and purchasing pages. You put in your products and design a few pages, and the checkout process is prescribed by the software product.

Working with one of these off-the-shelf pieces of software, you basically follow the path that the product lays out for you. The less you change the way a packaged software product "wants to work," the less risk you have.

Thinking You're Editing When You're Actually Creating

At the start of a software project, many people sign right up for off-the-shelf software. "Of course I want minimal customization! Who wouldn't want to reduce risk? That WordPress site template will work just fine."

It's important to understand, however, that choosing to edit something that exists over creating a new thing is like being on a weight-loss program. You are going to be on a diet for a very long time. You will be faced with a thousand temptations along the way, each one pushing you in the direction of more customization.

"Can't we move the button? Why not? Well, how much will it actually take? What about changing the order of billing address and shipping address? Why can't the page look the way I want it to? I spent a lot of money on a designer." And the absolute worst: "Wait. That built-in feature absolutely will not work for our customers. They must have it another way."

Sometimes at this stage, businesspeople say, "I was promised this out-of-the-box product could meet all our needs!"

Here's a rule: The out-of-the-box product rarely lives up to the "meets all our needs" standard.

I have seen many people start out intending to use the packaged software as it comes, straight "out of the box," only to get dragged into modification after modification as they confront reason after reason—some legitimate, some not—that the out-of-the-box system must change.

Choosing editing over creating is not just a matter of willpower, as our dieting metaphor might suggest, although willpower is unquestionably involved. Business requirements must be very well understood to make sure the product you pick is a good match for business needs. This means more time spent in planning and discovery to figure out whether some business requirement is going to be a major "gotcha" that sinks the whole packaged software plan.

Common Challenges with Off-the-Shelf Software

There are several common challenges in trying to use off-the-shelf software. I see these in business after business. It's good to be aware of them so discussions about project approach can be informed by others' experience.

Business Compromise

As suggested earlier, off-the-shelf software usually requires compromise on the part of the business to accept the way the software works. Sometimes those compromises can be significant. Business processes—the way people do their actual jobs—may need to change in order for the software to be deployed. Sometimes features will be given up entirely. Take the e-commerce example. Say the e-commerce software does not have the kind of discount code you want or a way of providing an electronic coupon. The sales team may have the use of this feature baked into their sales projections. Without it, they may not meet their sales goals.

The possible business sacrifices involved in an off-the-shelf choice must be identified. It's critical that all of the stakeholders on the business side understand the trade-off of customization versus risk.

Discovering You Made the Wrong Choice with Packaged Software

One of the worst situations you can get yourself into is to choose off-the-shelf software and realize at a late stage it would have been better to go the custom route. An all-too-common lament I hear is, "With so many customizations, this project would have been much easier if we just coded it from scratch." Easier *and* cheaper, because with too much customization, your programmer can spend no end of effort trying to change an out-of-the-box product to make it do something it was never intended to do. In fact, you might be surprised (and horrified!) to know how often it's necessary to pull the emergency brake, throw it all out, and begin again from scratch.

In short: "We want to avoid custom software and choose a packaged product" can be very easy to say up front and very difficult to stick to as the rubber meets the road. Be realistic and vet out your business requirements thoroughly against the pre-fab solution.

Breaking the Upgrade Path

If you find yourself doing too many customizations, the situation described previously, you are in danger of something called "breaking the upgrade path." All of us who use computers are familiar with software updates. They are those little reminders in your tray that you must download and install the latest version of Adobe. You click, it runs, and you've upgraded. The reason software gets upgraded is to fix bugs, introduce new functionality, and, more important, patch security vulnerabilities that put your computer and all its information at risk. Bottom line: You need upgrades and should do them on time!

Packaged software also has upgrades. Usually you'll hear this expressed as new "version releases." If you have customized the heck out of your packaged software, twisting the underlying functions to suit your desires and creating new functions for your own uses, you'll likely have a problem upgrading to the newest version. The most common thing that happens is all or most of your customizations will break and you will need to do them all over again. And then on the next release ... you get the idea.

Makers of packaged software expect that you will configure the product to meet your needs. But if you cross the line into a high degree of customization, you will be off their upgrade path. At some point the maker of the software will tell you they no longer support the version you are on.

Locked into a Partnership and the Product Roadmap

Think about the latest iPhone you bought. It had features and functions the last one didn't have. Who decided on these? Apple, of course! Which features and functions to release and when to do so is Apple's decision, based on customer input and the goals of Apple Computer. This is called the "product roadmap."

So it goes with packaged software. The company makes decisions on what new features and functions of Salesforce, Adobe Experience Manager, Quick-Books, NetSuite (and so on) are needed by the majority of customers.

What if your business is not in the majority? You need a particular feature because of a quirk in your logistics process. If it's not on the software maker's product roadmap, you must do your own customization.

Another tricky example happens when, because of your business needs, you want to integrate your packaged software with another piece of software. You might want a supplier to be able to send data to your system automatically through something called an "API." The API lets the two systems talk to one another. If the packaged software maker has not made APIs available for their product, you will not be able to create the integration with your supplier.

In other words, with packaged software you are in a business relationship with the maker of that software and dependent on the software roadmap. The smaller and more niche-oriented either your business or the software maker is, the more problems involving roadmap crop up.

Expense of Off-the-Shelf

One of the promises of off-the-shelf software is that it will be less expensive to implement and maintain than custom software.

As off-the-shelf products have grown more robust and complex, they have also become more and more expensive to implement. Many companies choose off-the-shelf software only to spend years of work and millions of dollars on specialized consultants to configure the software for their needs. You generally do avoid heart surgery–style risks with off-the-shelf software; but this is not always the case, especially if your particular configuration is something unusual. Often you end up needing the specialized consultants long into the future for support, upgrades, and maintenance.

As you might expect, the expense of implementing packaged software depends largely on who you are, how complicated the off-the-shelf software is to roll out, and your ongoing needs.

Where Packaged Software Works Well

Here are the conditions that usually need to be present in order for a packaged software choice to work well:

- The software is a common choice for your industry. You hardly know anyone in your business who doesn't use it. This means that if you have problems, everyone is having the same problems, and you will not be at a competitive disadvantage.
- You limit customizations when implementing the software. This is called "sticking to baseline." You work closely with your "integrator," the person or team who installs and configures the software, to make sure that any customizations will not break on the upgrade path.
- Your business requirements are well understood, and you have matched them to the software's functions.
- The likely product roadmap meets your needs.
- The packaged software is in the "mainstream," such as in the case of accounting software and CRM software.

Frameworks and the Blurring Worlds of Custom and Packaged Software

The crisp distinction between off-the-shelf and custom was very clear a decade ago. Now the picture is blurring.

As noted earlier, many off-the-shelf products have become much more complex. As they evolve, many of them are less like "out-of-the-box solutions" and more like platforms to work with. An example comes from the world of toys. If you buy Barbie's Dream Castle, it's all built and you're ready to play! If you buy a set of Legos, you have to build the castle before the dolls can have fun. With many off-the-shelf products, nowadays you're buying the box of Legos.

The newer off-the-shelf products are powerful and flexible. But don't be deluded that you can always hit "set up" and go.

At the same time packaged software is becoming more complex, the world of so-called custom coding is also changing. These days, it's unusual for programmers to code from scratch. Most are working with newer rapid-application-development "frameworks" such as Ruby on Rails, Python, and .NET that didn't exist fifteen years ago. If the word "framework" reminds you of the Lego example, you're right! Custom tools and packaged software are starting to look more and more alike.

The new coding tools and frameworks have created an environment where the cost and benefit of custom software has, in many cases, come back into line with off-the-shelf. It may cost $75,000 to implement off-the-shelf software and $75,000 to develop a custom application. In the case of custom, you get precisely what you want with no compromises.

Integrations vs. One Tool for the Job

To review, an "integration point" is a place where one system exchanges data with another system. For example, let's say your company has a website on WordPress and uses MailChimp for its e-mail program. You may have a sign-up form on your website where you collect the e-mail addresses of people who want to be on your company's mailing list. These addresses will somehow need to go from the WordPress site into MailChimp in order for that e-mail program to have the addresses. This is an example of a very simple integration point. Integration points are always risky, and we will talk about them in more detail later.

Because people know that integrations increase the difficulty of a project, they will look to avoid the risk by getting a single product that does everything. Therefore, no integrations will be required. There are many CRM systems that have website building and e-mail sending tools built in. Net-Suite is a good example of a "one product" solution. This product is called an ERP (Enterprise Resource Planning) because it handles an extensive set of business functions, including accounting, inventory, and logistics, and adds website building, e-mail, CRM, and so on.

But, you may not want to change e-mail programs just so you can work with a new accounting package. The time spent transferring everyone's mail over to the new program, not to mention the additional time lost as employees learn the new system, may be too much. Also, it is often true that a product that does many things can be a jack-of-all-trades and master of none.

Another thing that often happens in our Moore's Law world is that some new product enters the marketplace and your business stakeholders want to take advantage of it. If the silver bullet of e-mail marketing has the world abuzz with its power and flexibility, but you are locked into the e-mail tool of a one-size-fits-all solution, your company may find itself at a competitive disadvantage.

It is helpful to map out all your potential integration points and get a sense of your risk tolerance when it comes to integration. Then take a look at a possible "one tool for the job" solution and the compromises it will entail. Weigh the risks against the compromises.

To Phase or Not to Phase

The size of your project is directly related to how risky it is: Generally speaking, the larger the project, the greater the risk. Smaller projects are inherently less risky than larger ones.

Many business stakeholders believe they have no choice in the size of the project they are undertaking. Business needs, board mandates, and competitive issues seem to establish project size at the outset. For example, if your board of directors issues a mandate that your company is doing a nationwide rollout of Oracle eBusiness Suite ... well, that pretty much defines a big project.

However, business managers generally have far more flexibility than they think. Even a large project can be broken down into smaller pieces. For example, perhaps a company's entire corporate website needs to be redone. It is possible to take a piece of the website—say the investor area—and launch that first, then move on to the next part. It may be a little messy to have one piece of your site looking different from the other. But let me assure you, being over budget and late is far messier.

Objections to a phased approach generally focus on the less-than-smooth user experience. In the previous example, users would come to the main website, click on the Investor tab, and end up on another website with different navigation and, perhaps, no easy way to click back to where they were.

In my experience, these negatives create a lot of fussing but are always far outweighed by the positives. It is extraordinarily beneficial to be able to roll out a piece of something, learn from that experience, and then do the next piece. On a human level, people—even very talented people—can only master so much depth and breadth at once. Many large software projects face problems just because of their sheer size and the inability of the people involved to master all the details.

Here are two ways to make a big project smaller:

- Split it up into pieces delivered one by one
- Reduce your audience size; release the software to a limited group of customers in a pilot, discussed next

HINT: BUT WHAT ABOUT THE COST?

It may initially seem that one-by-one deliverables raise your costs. And this approach will certainly extend your timeline. However, when you honestly weigh your risks and assess the likelihood that your large project will be late and over budget, this delta may evaporate.

Bigger Is Not Always Better

Emotions can also be an obstacle when discussing project size. A piecemeal approach can feel deflating to stakeholders, CEOs, and boards. People much prefer the idea of a big reveal: "Ta-da!" The switch is flipped, and the new logistics system is online. If the end product is consumer facing, marketing folks want the opportunity to create excitement.

There are cases, of course, where the "big bang" approach is necessary. For example, if you are switching your point-of-sale cash register system, it must be done in one fell swoop. However, aside from these few instances, most projects can be broken into chunks, and the internal resistance you'll face centers around the eagerness to achieve the "ta-da." Many CEOs and boards, however, have not actually had the experience of a large-scale failure. Those who have will immediately see the value of doing a project in smaller phases.

The Pilot Approach

"Piloting" is one way to phase a project, but with a slightly different twist. Piloting means launching some small version of the end product. "Small" means one or both of the following conditions: (1) with a reduced feature set, and (2) to a limited group of people.

The pilot approach virtually ensures you will learn things you didn't know. Even in businesses with deep customer knowledge, pilot projects surprise stakeholders with learnings about what end customers really want.

Let's say you want to launch a new tool on your website. It's a "manage my account" page where users can select settings and change the way the website looks.

In a pilot, you might select a small group of customers and engage them in the design process. Their input is collected and plans are made. When the end product is launched, often to a slightly larger group, the learnings are incorporated into the full rollout. You get more efficient at putting together the "Ikea desks," and you discover if you have any lurking "heart surgeries" you didn't know about.

In my experience, companies with a "pilot culture" develop more successful software products than those without.

Why Not Pilot?

Seems simple, right? Define a pilot, execute the pilot, learn from the pilot, and then roll out the final product. Why wouldn't everybody do that?

The truth is businesses simply don't. Mostly it comes down to impatience. People want their products done now! By the time a project gains approval, requirements are collected, and teams are assembled, many businesspeople are already shocked at how long it's taken. Inserting a pilot phase will often take an additional six months. You need time to launch the pilot project and collect reactions from it. Besides. We know what we want. Why don't we just do it!

Also, it often looks as if a pilot will add cost, increasing the budget by 10 to 15 percent. The truth is, pilots invariably save money in the end. Features that were considered critical may be eliminated entirely as users greet them with so-so reactions.

CHAPTER 4

Teams and Team Roles and Responsibilities Defined

Film director Robert Altman famously said that 90 percent of a director's job is casting. The same is true for software development. Getting the right team in place is, without question, the most important thing you can do to ensure the success of a software development project. But wait! You may ask, "How do I build a team when I don't even know the technology I'm choosing? Don't my team members have to be familiar with the technology?

The answer to all these questions is "not necessarily." First, let's discuss teams.

Teams and the Roles on Teams

Over the years, I have noticed that businesses tend to be unfamiliar with the roles necessary on a software project. There are lots of blurred lines, mixed up roles, missing roles, double teaming roles, and so forth. It's like a soccer game where you're not even clear who's on offense and who's on defense, never mind identifying the goalie.

Without a clear understanding of roles, it's impossible to staff a project. Whom are we looking for, and what holes do we need to plug? Sticking with the sports metaphor, it would be like an NFL team owner going into the draft meeting without a clear understanding of what a cornerback does.

Of course, the members and designated roles on a software development project are more flexible than those of the NFL and can vary greatly depending on the project. Here, however, I'm going to discuss the major roles and give you advice on how to fill them.

A common comment when people learn all the roles is, "Wow! There are a lot of them!" Yes, there are. Keep in mind a note from the Introduction. I am speaking here about medium- to large-sized software projects. On smaller projects, it's common for one team member to wear multiple hats. How and whether to blend is directly related to the size of the project and to the talents of your team members. In a small project, it is of course appropriate for team members to do several jobs—more on this later. The larger the project, however, the more distinct the roles need to be.

Still, the structure is the structure. Anyone tackling a software project must be aware of the basic roles, just as anyone entering a soccer game must know about the standard game positions. Once you understand that, you can make decisions on your own.

All projects have the following:

1. A project leadership team
2. A project execution team
3. Business stakeholders, also called "internal customers"

Project Leadership

The project leadership can be understood as a bridge between the "techies" and the business. As a bridge, the project leadership team will have technology folks and businesspeople (Figure 4.1).

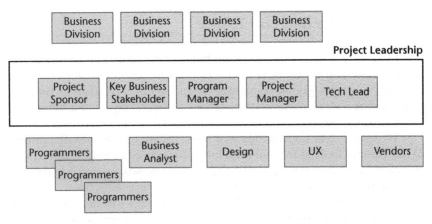

Figure 4.1

The Key Business Stakeholder

There may be many internal customers for a software project. But one of them is usually most important and becomes part of the project leadership. In cases where there are many internal stakeholders, this person serves as the "voice" for all of them. For example, the key business stakeholder might be the head of finance who knows the company needs a new accounting system. It might be the head of sales who is pushing for a new CRM system. It might be the head of logistics who wants a new method for tracking inventory. Or it might be the head of marketing who wants a new website.

In any of these cases, it's usually easy to identify the key business stakeholder. She is the person raising her hand and saying, "Hey, we need this!"

As noted earlier, sometimes there is more than one key business stakeholder; but more often than not, there is one main key business stakeholder.

The Project Sponsor

This is the business manager charged with overseeing the software development project and with the accountability for its success or failure. This is the person in the business with responsibility for the project. Top leadership has come to this person and handed them the directive to make sure this project gets done. This person is often the chief information officer, chief digital officer, or even the chief operations officer. A need has been identified, a project must be started, and the project sponsor is on the hook for getting it done. All the teams report to this person who, in turn, interfaces with the rest of the organization.

The project sponsor's role is as key project decision maker—she is, as the saying goes, where the buck stops. Another role is as a buffer. It's largely the project sponsor's job to provide the bridge between the project and the rest of the organization, and to act as a buffer between the organization and the technologists. Without this buffer, the technologists are going to be faced with too many clamoring voices from the business side, all stating their various (and sometimes conflicting) needs, wants, and desires.

In essence, the project sponsor becomes the negotiator-in-chief. For example, when the technology team says a business requirement is too complex and time consuming, but the business stakeholders say it is a mission-critical piece of functionality, the project sponsor will need to mediate the solution.

Hint #1: The project sponsor and key business stakeholder are sometimes the same person. Say, for example, the CIO says the company should transition all servers to the Cloud so that he can better scale resources across the business. In this case the CIO is both the main internal customer and likely to be the project sponsor.

Hint #2: The project sponsor has a very close interlock with the program manager, and they often split responsibilities.

The Program Manager

People familiar with advertising agencies may be more comfortable with the term "account manager." The program manager role is usually seen on medium to large projects. This is a person experienced in running large software projects who organizes all the teams, calls the meetings, gives directions, identifies needs, and leads all the other team members through the processes (we will discuss) of deciding on technology, picking team members and vendors, gathering business requirements, breaking down the project into tracks, budgeting, and leading execution. If the project sponsor is nontechnical or inexperienced in the type of technology under consideration, the program manager will serve as her trusted technology partner and guide through the process. This is the role I most often play on projects. So, in other words, it's the kind of person who knows all the information in this book!

If you are getting the sense that this person is a kind of general in charge of forming and executing a battle plan, you're right! She is! She would be the one to lead the business through topics like project approach covered in Chapter 3.

The program manager is often an outsourced role, but sometimes an internal resource fills this role. We'll talk about using internal vs. external staffing in a separate section of this chapter.

The project sponsor is often a businessperson (e.g., chief operating officer), and the program manager is often a senior technologist with years of experience rolling out software projects and is comfortable with two or more project managers working under her.

Project Manager

You might be thinking, "Wow, what a lot of similar sounding job titles! What's the difference between the program manager and the project manager?" You are right. The titles in software projects seem almost designed to confuse. Nevertheless, we've got them, so we have to work with them.

Figure 4.2

If you find yourself getting confused between program manager and project manager, think of the program manager as the ship's captain and the project manager as first mate. Figure 4.2 may help in clarifying roles.

The project manager (PM) is a level down from the program manager. He or she is formally trained in the methods of project management. He is in charge of interacting daily with the technical team, making and tracking the detailed plans, and using project management tools to report on progress. You will see this person conducting daily status meetings, preparing agendas, capturing to-dos and follow-ups, and making sure they get done. This person will have mastery over project management tools such as specialized spreadsheets with lots of macros, ticketing tools, and project planning tools such as Microsoft Project or Jira Portfolio. Frequently the PM will have a project management certification called a PMP.

A good PM is worth his or her weight in gold, and no software development project can move forward without one. The PM will know how many hours have been assigned to program a certain widget, and if the task is on schedule.

Multiple Project Managers

If a software development project involves one or more vendors, there needs to be a PM at each of the vendors. For example, on a software development project with three vendors, ideally there would be four PMs. The business itself must have one, outsourced or on-staff, and there should be one at each vendor. When you start off with a vendor, make sure to ask them who the PM will be. This is a role that often gets left out, but is critically important. If you have a lead or major player vendor on the project, their PM may, in certain circumstances, serve as your PM.

One big mistake people often make in the selection of a software development project manager is in thinking they need a PM who understands their particular industry. For example, in media companies, the trend is often toward selecting PMs who understand the terminology such as "headlines," "bylines," "slugs," and so forth. If a PM isn't facile with the vocabulary, confidence is immediately lost.

And yet, nothing could be farther from the ideal strategy for selecting a PM. If you are engaging in a software development project, you want a PM who excels in project management for *software*. This is an invariable rule.

HINT: SOME THINGS ARE EASIER TO LEARN THAN OTHERS

A software-savvy PM can learn information necessary to a specific industry far more easily than an industry-savvy individual can be taught the project management skills necessary for a successful software project.

Confusion About the Project Manager Role; It's More Limited than You Think

This is a book about project management, so it's important to get the vocabulary straight. To say someone is the "project manager" seems utterly comprehensive in scope. And yet "project management" is a defined career track, much narrower than the title implies to most people. A certification in project management is dedicated to the specific principles, tools, and methods required to run a project. A project manager generally does not set strategy, nor is this person expected to guide business leaders in decisions on high-level project approach.

Many books about project management further add to the confusion and blurring of roles. Topics covered for project managers are more appropriate for project sponsors, such as building a business case for a project or calculating future ROI. While it is clearly critical for project managers to know about all these subjects, it's also critical for everyone to know what's in his job scope and what's out. We'll talk about boundaries and "driving in your lane" later in this chapter.

Project Team

After the project leadership, we have the actual project team. These are the people with the "hands on the keyboards," so to speak: gathering requirements, analyzing business needs, defining tasks, and doing tasks.

The Business Analyst

The business analyst (BA) is the person who collects, digests, and codifies all the software development requirements from the business stakeholders. This person will often be found interviewing members of different departments, learning from them how the software will meet their needs and what it is expected to do when it's done. The BA then writes all this up in a way that both the programming team and businesspeople can understand, with all the necessary level of detail, including all the diagrams and descriptions programmers will require.

Oftentimes people think the BA and the PM are the same person or try to blend these roles. Later in this chapter, I will discuss how to blend roles. But in general, I have found the BA/PM is not a good combination. For one thing, the PM has enough to do. For another, the BA is often more of a "big picture thinker" whereas the PM must be quite detail oriented.

User Experience

The user experience (UX) person is a professional who designs the user interaction (i.e., which clicks lead to what) and navigation in a piece of software. This person will often be found developing sitemaps and wireframes (mockups that focus on functionality). Many times a UX person develops interactive wireframes that provide some level of click response, so you can see how the software will behave. Sometimes the UX person can often also do visual design. Sometimes the reverse happens—a visual design person also does the UX.

Designer

Pretty much everyone knows what a designer is, and this role requires little explanation. However, for software development projects, it's especially important for the designer to be integrated into the team and to consider blended roles such as user experience and design.

It is much better to avoid designers who come from a primarily (or exclusively) analog world, meaning those who have done most of their design for print. Ideally, your designer should have a fundamental understanding of the technologies being used and what is possible within that technology.

The Programmers

Broadly speaking, any large piece of software can be described as having a front end and a back end: The front end is the "interface," or the part of the software that users see and experience while the back end is the "processor" that uses the information collected by the front end. Consequently, your programming staff will probably break down into two levels of engineers: front-end and back-end programmers.

It's important to understand that front-end programmers are not user experience people or designers. What they do is take the user experience and design material and write the code that translates it into the actual front-end screens. They're often working with technologies called CSS, Javascript, and HTML.

Back-end programmers, on the other hand, are often database programmers. They are in charge of putting in place the systems that will feed data into the front end and take data from the front end into the back end where it's processed and stored.

Architect

A common problem in software development projects occurs at the end, when the software is deployed on the actual machines where it runs. Maybe there are integrations and different systems that need to talk to each other. Maybe the site has to support tons of traffic and lots of simultaneous users.

Conflicts arise when something isn't working. Often, programmers blame the hardware people for not having enough machines or the right machines, while the hardware people blame the software people for writing code that doesn't run well on the machines they have. We will talk about this situation in later chapters that discuss the challenges that come up around launching a software development project. But right now, we're talking about roles.

To avoid the conflict I just described, it's important to have a person— often called an architect—as the go-between for the programmers and the managers of the hardware. This person will be accountable for developing what's called an "architectural map" or "footprint." This architectural map will identify the number of machines, types of machines, and size of machines necessary to run the software. The architect will work with the systems administrator to set things up so that integrations can happen.

Systems Administrator

Obviously the software will need to be deployed on machines that will run it. There must be a person in charge of this hosting environment. That's called the systems administrator or Sys Admin (Figure 4.3).

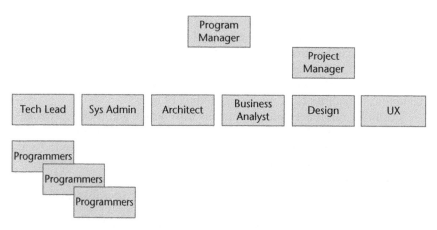

Figure 4.3

Team Member Choice and Blending Roles

Previously, I've outlined the key roles found in software development projects. With that list you will have a starting point to get all the bases covered. As you plan out your team, ask questions like:

- Who is going to keep track of tasks, schedules, and budgets? (the PM role)
- Who is going to collect business requirements and document them? (the BA role)
- Who will give us the wireframes showing us how the software will behave and interact? (the UX role)

Getting All the Roles Covered

Looking at this list of roles can be alarming. The number of people involved can be enough to make you think your budget is blown before you start. And it's true: In very large software projects there may be quite a large team, with several people playing each role. But in smaller projects, these roles are frequently blended.

However, many people inappropriately blend roles without even knowing they're doing it. This often leads to trouble. Assuming a particular task is part of a person's job when it's not leads to conflicts with staff and vendors.

For example, a member of a programming team may find, midway through the project, that his boss expected him to be on top of the interactive user experience, but that wasn't his expectation at all. A businessperson may think, "Well, this programmer is working on the front end, right? That means doing all the interactions that a user will experience." Again, this is a common, sometimes fatal, misunderstanding of roles.

A similar misunderstanding happens with design and user experience. A visual designer may have little or no background in user interactions; yet in many software development projects, user experience is commonly handed over to designers, with negative results.

Then there are the missing roles. I've seen vendors make certain requests of the project manager only to discover there *is no* project manager, because the client expected them to provide that service.

All these examples spring from a lack of understanding of the roles necessary to execute a software development project. And they can lead to budget and schedule overruns, as well as conflicts and discomfort.

My point is that even if you blend roles, you'll need to remember that all of the key roles *must exist at some level*. The critical skillsets must get covered so there are no gaps. All the jobs must be done. Only once you are clear on what's needed, then you start to mix and match.

As mentioned earlier, the sole unbreakable rule when it comes to blending roles is not to blend the PM (project manager) role with anything else. I have found this to be true not only because of the number of things that the PM has to do, but also because of skillset and personality reasons. The PM stands alone.

Caveat: The old rule that "you get what you pay for" tends to be true here. A specialist is always better than someone who has extended his or her role into another area.

Real-World Examples for Role-Blending

What follows are some real-world examples where role-blending works well. See also Table 4.1.

Project Sponsor as Program Manager

Let's say the project sponsor is a real gem. She may have even worked on a software project before. Further, she has cleared a good chunk of her time to lead kickoff meetings, be present at weekly status meetings, help the business

Table 4.1   Role Checklist and Blendability Worksheet

Role	Blend with?	Notes
Project Sponsor	Program Manager	A Project Sponsor with software development experience can take on the Program Manager role.
Program manager	Business analyst	CIO, CTO, or consultant
Project manager	Do not blend	Detailed thinker. Choose PM with good software development background as opposed to industry-specific.
Business analyst	UX and program management	BAs are big-picture thinkers. A visually trained BA can do UX. Sometimes a vendor's account manager will also serve as a BA.
User experience	BA or front-end programmer	A visual BA can also serve in the UX role. Also, a front-end programmer with good communication skills can play the UX role.
Designer	UX	A designer who is very experienced digitally can sometimes make the transition to UX.
Front-end programmer	UX	If the front-end programmer has good communication skills, he or she can sometimes do UX tasks.
Back-end programmer	Architect	A good back-end programmer may be able to recommend systems architecture.
Architect	Back-end programming	A good back-end programmer can play this role.
Systems administrator	Architect	A good systems administrator can also play the architect role.

analyst shake down the business stakeholders for requirements, and respond to red flags raised by the PM.

In this case, the project sponsor may also serve as the program manager.

Program Manager as Business Analyst

Another common example is for the program manager to serve as the business analyst. Often, the program manager has come up through the business analysis track. She certainly has to do a lot of analysis in her job! In this case, she may be the one to write the business documentation and take over the BA role.

Front-End Programmer as User Experience

Some front-end programmers with good communication skills and experience in user interaction can sometimes do user experience.

Design, UX, and Business Analysis

These three roles (design, user experience, and business analysis) can some-times be found in one person. It can be appropriate to blend these roles, especially on small projects. Indeed, I myself have performed all of these tasks on certain software projects because I have a background in business analysis, user experience, and design—and I'm also a senior technologist. In terms of cost-efficiency, this is obviously an ideal situation.

Back-End Programmer as Architect

A good programmer (especially a good back-end programmer) can often play the architect role. But be careful here: Not all programmers understand the hardware side. In fact, it's becoming less common these days, as technology trends distance programmers farther and farther away from the hardware.

It's been my experience that programmers who are more seasoned (often, this literally means older) have greater experience on the hardware side and can interact more effectively with a hosting company.

SysAdmin as Architect

Similarly, a very experienced systems administrator can serve as an archi-tect and lay out a systems footprint; this is especially true for small- and medium-sized projects. For larger projects, it is helpful to have a very senior technologist serving as the architect. That person will have a thorough understanding of the interplay between hardware and software as well as networking and integration issues.

Professionals and Personalities

It can help a great deal to have insight into the kinds of professionals who work on a software development project, especially if you've never worked with these professionals and specialties before. It can also help in hiring to know what traits you are and are not looking for. Understanding personalities is also essential if you want to blend roles.

Programmers

Programmers are an interesting bunch. The best programmers are highly focused, extremely detail-oriented, and thorough. Programmers also tend to be black-and-white thinkers and are very literal-minded. They prefer to follow specific instructions with a minimum of vagueness because this increases their ability to achieve success.

Programmers can show impatience with big-picture thinkers who like to stay at the 30,000-foot view without much consideration for what it will take to achieve the vision. Programmers can be crotchety or even dismissive of such "hand wavers"—to their way of thinking—who unfortunately may be senior, or even C-level, executives.

Programmers are usually not big talkers and generally don't like to participate in meetings having to do with big-picture strategy. They prefer to stay focused on the pragmatic: what needs to get done and what's the most efficient way to do it. A programmer's greatest frustration is shifting or unclear business requirements.

Programmers are the most mission-critical team members in a software development project and the hardest to hire. This is because few of us have the wherewithal to evaluate a programmer's skillset. I have found that even other good programmers fall short on this score. One key reason is that one programmer is rarely up to another's snuff. Further, programmers tend to work in isolation, using the methods and styles they've found to be effective for them.

There are indeed highly individualized aspects to the work, which makes it hard for programmers to evaluate one another. However, there are ways to find and keep good programmers. These generally involve:

- Testing programmers using programming tasks and code review
- Putting a premium on the qualities of focus, thoroughness, and accuracy
- Deemphasizing nonessential qualities (often important for other roles) such as personality, personable-ness, and good communication skills

With this frank discussion of the traits found in many programmers, it will be evident that asking a programmer to do big-picture tasks such as business analysis and user experience can be a bad idea.

Project Managers

Good project managers are, by definition, detail-oriented. This is the key trait you're looking for. A good project manager is not intimidated by a blizzard of tasks, budget numbers, and scheduling details: He quickly sweeps them up into spreadsheets and timelines. He loves to put systems into place that give the stakeholders insight into how the project is doing.

With all this said, it may surprise you to learn that most project managers are also *not* big-picture thinkers. You may ask, "Isn't that what they do after all—keep track of the big picture?"

Based on this misunderstanding, many managers ask the project manager, for example, to do business analysis or run meetings in a kind of account manager role. But how can a project manager run a meeting when his or her role is to take notes, keep track of what's decided in the meeting, and who's assigned to what task? Moreover, just because a person has a skill in organizing and tracking the details does not necessarily mean he or she has a grasp of the big picture. It's a forest-for-the-trees kind of thing.

I frequently overrule my wonderful project managers. Looking at the reports and the detail, I will say something like, "I know the programming team seems on schedule, but they look nervous to me. There's something going on with those guys and we need to check into it."

In short, project managers can be black-and-white thinkers, which is the key reason many find it challenging to do business analysis. The traits you're looking for in a good PM are level-headedness, detail orientation, and facility with numbers, spreadsheets, and planning tools.

Be careful to avoid project managers who are too in love with the tools, or—the worst common quality among project managers—are overly rigid. A plan is just a plan. It needs to have a certain amount of elasticity to accommodate the growths and shifts that will inevitably occur along the way. A good PM is flexible in addition to being detail-oriented.

Reminder: It is not advisable to blend the PM with any other role.

Business Analysts and User Experience People

BAs and user experience people are much easier to evaluate and hire. They are the big-picture thinkers in the project and therefore have much more in common with the business leaders of the project.

The key qualities here are excellent communication and listening skills, as well as the ability to create stellar documentation that will allow both technologists and nontechnologists to visualize and understand the software development project as it unfolds. The key tools these professionals use will be wireframing software, PowerPoint, Microsoft Word, and Visio.

Architects and Systems Administrators

Architects and Systems Administrators tend to share the same personality traits and skillsets as programmers. Many times architects are often professionals who have graduated into this role from a career in programming.

Insource or Outsource: Whether to Staff Roles with Internal People or Get Outside Help

In my experience, this is an area where tempers get hot and opinions are definite. The CIO insists on an "in-house programming team" because "you don't want to be at the mercy of a vendor." Or, the CFO declares that the budget will not support the salary and benefits of two project managers. Assertions are hurled around about whose perspective is the best business strategy in terms of efficacy and money. In the end, the matter is decided based on whose opinion has the most weight in the organization.

Following are some facts that will untangle the issues and help sort out the best strategy for your organization.

The Myth that Insourcing Programming Is Better

Many companies undertaking a software project have heard about other companies who "spent a ton of money on external software developers" and make the judgment that it's better to go with an in-house team to execute the project. These in-house guys don't charge by the hour. They are a fixed, known cost. Feelings often tilt in this direction if there is an existing team in place, even if that team is just one or two people.

The in-house programmers naturally support this view. They set forth all kinds of information to management underscoring why in-house is the way to go. It's amazing to me how many business leaders take these opinions on face value without seeing the obvious, and perfectly understandable, self-interest at work here.

Often the in-house team is super-excited by this new software project. In the normal course of business, their jobs may have been reduced to fixing bugs on existing systems with minor enhancements to the system as the only bright spots in their days. Who wouldn't want the chance to have their hands in a shiny, new software project? In addition to the excitement, it's a chance to expand their ranks (previously under-resourced) and learn new skills. It's very common for a one- or two-person in-house technology team to add four or five members to tackle a software project. Again, who wouldn't want their department to receive new resources?

And here's where the problems come in.

Many businesspeople mistakenly think that programmers are magicians who can master new technologies and programming languages at the drop of a hat. This is simply not so. Remember the conversation about the expense of in-house versus outsourced? Retraining the internal team costs money. And it takes time. Of course, your new hires will have the requisite skills. But the legacy staff might not. Further, when the internal team is trained, they have new opportunities and new horizons. It's not uncommon for programmers to jump ship mid-project. Their career prospects have brightened, but your project still needs to get done. The project grinds to a halt as programmers are rehired.

Another factor with using an in-house team is that their skills often define the technology choice. We'll get into technology choice in Chapter 5. For now, imagine you have a team of .NET programmers. When it comes to rolling out a new shopping cart, they will only look at .NET solutions. What if an open source cart is a better solution for the business? It's possible the internal team won't even put that on the table to consider.

Inexperience with Projects

Internal teams may be very good at keeping technology running in the normal course of business. They know your systems and put out fires with gusto. However, they may not have experience working on a large software project with a defined kickoff, development phase, and go-live. They may never have experienced such a thing before. Again, this may be hard for business managers to understand. They may think that all programmers have the launch of Twitter somewhere in their DNA. But it's simply not true.

Launching software projects is very different than business-as-usual, even if your business-as-usual is quite pressure-filled.

How Knowledge Goes Stale

Another challenge with internal teams and software development is their lack of exposure to the broader world of software development. Remember Moore's Law? It's at work on them, too. For every 18 months they are inside your company, the software world is advancing. To be sure, many internal software developers read voraciously, keeping up on the latest trends and making sure they don't fall behind. They may, and often do, have small side projects to keep their skills honed. It still may not be enough to keep them competitive with an external team.

Outsourced Teams

Based on the previous discussion, it's easy to see the advantage of outsourced teams. They do have "launching software" in their DNA. They are working in dozens of different companies, deploying software. So they bring the latest learning not only when it comes to technology, but also about what has succeeded and failed in the broader world and why. Their skills are constantly sharpened. And they are well versed in the phases of a software project in contrast to business-as-usual.

An outsourced team comes in, does what you need, and, in a well-run project, leaves on time.

When to Use Internal or External Teams

As a general rule, an internal tech team is a good idea when technology or a technology product is the core of your business. An external team is a good idea when your business is to deliver a product with technology in a supporting role.

It's not as easy as you think to decide if you are a technology company or not. First, technology has become so deeply embedded in businesses that a businessperson may not be able to tell where the technology ends and the product begins. Further, technology is so sexy to many businesspeople that they delude themselves about where it fits into their businesses.

Ask yourself the following questions: Do you make a technology product? Has the digital revolution changed your business so much that you are indeed delivering technology rather than physical goods? Is technology a core competitive advantage for you? Or are you a product manufacturer with technology in a supporting role?

Here are some examples. A consumer goods company makes a food product. These days, they need lots of websites to deliver everything from kids' games to nutrition information in multiple languages. Still, the fact exists this company makes *food*, not websites. The best strategy here is to hire great external developers.

A media company must take all its magazine titles online. Here the question is more complicated. The digital revolution has so changed media, you could argue technology is core to the product, which is no longer a print magazine, per se, but rather "content," a fluid concept deliverable across printed pages, websites, and mobile phones. As you might imagine, many media companies have indeed decided to invest in internal programming and project management teams. And many have not.

A startup company has decided to provide technology to pharmaceutical companies by helping them to manage drug trials more efficiently. Clearly, in this case, technology is core to the mission, and an internal team makes sense.

A final consideration in the internal vs. external debate has to do with the nature of your project. Many software projects are special events. A board of directors or department head has approved new budget dollars to do something unusual and outside the normal course of business. In this case, it usually makes much more sense to work with an external team.

I often say, "Large software projects have the gravitational force of the planet Jupiter." As they unfold and move through the organization, they actually transform the organization. Needs change. People change. Staff members jump ship. Excellent members of vendor teams approach the organization and say, "Hey, do you think I could work here?" Said simply, you may not recognize your own company when the project is done, especially if the project is large. For this reason, too, you may not want to make long-term staffing decisions until the project is over and you see what your needs are at that time.

Roles Easiest to Outsource

If you choose the outsourcing route, certain roles make most sense.

- *Program manager:* This is a very specialized role. This person's job is to run large software projects. Unless your organization has technology at its core with large software projects happening constantly, get an external resource for this role.
- *Design/UX:* This one's obvious. Even the most technological companies may outsource these roles.
- *Programming team:* As discussed earlier, it often makes most sense to outsource this team.

Roles "in the Middle"

These roles may make sense to keep in-house, depending on your business.

- *Project manager:* If you have lots of projects, and many organizations do, it can make sense to have a good project manager on staff. If a large software project comes your way, this person may still need support from outside resources experienced in large projects.
- *Business analyst:* This person's job is to capture what the business wants in the software project. Obviously, someone inside an organization will know the ins and outs of the business best and will be best at capturing nuances. A caveat here is the business analyst has to have some understanding of software. Her job is to capture requirements in a way that a programming team can understand.

Roles that Are Usually Internal

- *SysAdmin:* The IT group in an organization will often supply this role.
- *Architect:* As with SysAdmin, often the IT group in an organization will supply this role, especially if the organization has an established infrastructure on which the software must run.

Vendors and Hiring External Resources

I have an adage: "The only vendor that works is a vendor that's worked before."

The value of a software vendor who brings a project in on time and on budget simply cannot be calculated. They are difficult to find, and when you have found one, it's important to remember them for the future.

What happens if you don't have one? You are going to need to go out and interview people you trust and ask them to recommend a good software vendor. Be careful to get recommendations from people who have an investment in *you*—not in the vendor. Ask yourself the following question: If the vendor fails, will the person who recommended them feel bad about the situation you're left in or will she feel bad for the vendor?

HINT: IT'S WHAT THEY DON'T SAY

It's been my experience in vendor selection (and, frankly, in employee selection) that references lie. One trick that I've developed is to listen for the things that people say are *not* true of the vendor. For example, if someone says the vendor is never late, I'm suspicious as to why they have brought up the topic of being late, and I will ask a lot of follow-up questions.

Some Tech-Types to Avoid: Dot Communists and Shamans

We have all heard inspiring technology success stories in the last two decades. There are many technology professionals who have been lucky enough to participate in exciting start-up companies.

As you make choices about hiring internally or partnering with a vendor, one thing to be on the lookout for is the species I call "dot communists." These are folks who have participated in technology start-ups (either a true start-up or an R&D effort inside a company), and it has literally gone to their heads. They are starstruck. Business leaders may be starstruck as well, as the dot communists' stories are often glamorous and exciting!

One of the dangers here is that a lot of the successes in the dot-com era (both back in the late 1990s and the new one happening now) have depended to a large degree on luck and timing. Also, dot-com companies are often considered successful if they are "hot" with a high valuation, even if they are losing millions of dollars a year. Many of these companies ultimately fail. Many are started and flip so quickly that it's hard to tell who is responsible for their "success." Needless to say, there is a lot of speculative behavior around these endeavors.

Technology professionals who come from this background can have a lot of ego and very little experience dealing with an actual P&L. Dot communists often apply for roles as program managers or even CIOs.

The Shamans

Programmers, software developers, and technology people in general can be an intimidating bunch. First, they have lots of specialized training in arcane corners of math and science that most of us find impenetrable. They can read and write computer code in a way that, to an observer, can almost seem like magic. And let's be honest: These engineering types may not be very talkative or forthcoming in interpersonal interactions. All of these factors, plus many others, conspire to create what I call "the shamans."

The word "shaman" refers to an individual who has the power to act as an intermediary between the natural and supernatural worlds. By definition, a shaman possesses powers that mere mortals do not. Unfortunately, many technologists and engineers act like shamans, hoarding their special magic and perpetuating the idea that they have a unique ability to work the alchemy that is technology.

Also unfortunately, many businesspeople perpetuate this culture by treating technologists as shamans. Mostly this occurs due to the fear and intimidation many people feel when confronting something they don't understand—especially when it has to do with technology. Sometimes it seems businesspeople even *enjoy* the romantic notion that technology is magical. It makes the day more interesting with all this wizardry going on.

Project and business leadership should not perpetuate the idea that technology is magical and therefore the province of only a few initiated sorcerers.

Nip this in the bud. Be alert for technologists who act like shamans. Do not treat technologists like shamans, and do not put up with a technologist who acts like one.

One way to dissolve the mist surrounding shamans is to ask questions, lots of questions. Everyone needs to remember that there are no stupid questions. A good technologist should be happy that his business stakeholder wants to thoroughly understand the project. Business stakeholders must pursue the answers they need, and if they don't understand the explanation, ask again. Yes, people get impatient with this process, but it is essential.

"Listening loudly" is a phrase a client of mine uses, and I absolutely love it. Listening loudly captures both the quality of active listening and the imperative to notice subtext in what you are hearing. I have seen people with absolutely no computer background make very sophisticated and, in my opinion, quite accurate technology decisions just by listening carefully.

Project leadership and business stakeholders can unmask shaman-like technologists. Tell shamans using too much jargon to stop. I have found that most (if not all) technology issues can be understood by the layperson. Yes, it takes time. But this investment is fundamental to a project's success. The alternative, as we'll see later when discussing business leaders' responsibilities, is to proceed throughout the project basing critical decisions on a murky understanding of what is going on.

Boundaries, Responsibilities, and Driving in Your Lane

In the beginning of a software project, it is critical that everyone understands what is included in his role and what is not.

As a general principle, business stakeholders are often unaware of their responsibilities and therefore shirk them. Team implementers, in contrast, often overreach.

Here's the rule: Business stakeholders get to decide, and the project team gets to recommend; or, more specifically:

- The business leadership and business stakeholders make decisions upon matters of technological strategy.
- The implementation team makes decisions upon tactics, the "how."
- If an implementation tactic impacts a feature, it's bumped up to a strategic matter. In other words, back to the business.

An analogy helps in understanding: Say a child has to undergo elective surgery and there are two possible options. The surgeon expresses his opinion in favor of Option A. The parents, for their own reasons, prefer Option B.

No matter how strong the surgeon feels about it (he may even decline to do the procedure), the decision is the parents'.

Once in the operating theater, the surgeon determines how to perform the procedure or deal with unforeseen things that come up.

Keep this analogy in mind when it comes to software development. In the next chapter, we're be diving into technology choice, where it's critical.

Techies Who Don't Drive in Their Lane

One of the points of discomfort in the previous example is that *the doctor* possesses all the specialized knowledge. Who are the parents to decide? They didn't go to medical school? Of course, such a question is ridiculous. They get to decide because *it's their child.*

In the beginning of a software project, you may hear programmers and other techies wail, "But it makes no sense! Clearly we need a new database. How can they decide? They don't even know what they're talking about!" This happens, by the way, whether the team is internal or external.

Unlike doctors, who receive training on these boundary-related matters in medical school, technologists don't.

Programmers are the biggest offenders on this matter. They leap into making strategic technological decisions without even realizing it's not theirs to decide. They may say things like, "We have to go with the .NET solution." Or, "The database must be replaced." Or, "Our new content management system is going to be Drupal." Programmers must be educated by experienced managers that their job is to recommend and educate, *not* to decide.

It's not enough for the project manager or program manager to know what decisions must get elevated to the business, though they clearly help. The implementation team, designers, UX people, business analysts, and programmers will be making a thousand decisions a day. In a team with a solid understanding of boundaries, you will often hear, "We came across this thing, and I think it involves a business decision."

Good project management must include the correct flow of information for decision making. Otherwise, inappropriate calls get made, or projects get stalled for lack of decisions.

Business Stakeholders Who Shirk Responsibilities

Driving outside her lane may not be the techie's fault. It may be a business stakeholder who's pushing her there.

Business stakeholders with little experience in technology often look at the tech team with wide eyes and palms raised to the sky and say, "I don't know? What should we do?" After a few interactions like this, the programmer may think, "It's my responsibility to decide."

Don't let this happen in your organization. As with the example of the parents, often the least technically expert people in the whole firm are responsible for the firm's technological strategy. This is especially true if the business is undertaking a large software project with significant budget and strategic implications.

It's the responsibility of the project leadership—the program manager, project manager, and (often) business analyst—to make sure the business stakeholders have all the information they need to make strategic decisions. But the business stakeholders must step up to their responsibilities.

Business Stakeholders, Step Up!

It is critical for business stakeholders and organizational leadership to accept that however unprepared they may feel, *they* are the strategic decision makers in this project. One of the things most fatal to the software development process is when the business leadership abdicates its responsibility due to a lack of confidence. Projects dissolve into debate and chaos. But with support and determination, nontechnologists can make good technology decisions without formal training. It's one of the main reasons I'm writing this book. Businesspeople can do it by asking the right questions and getting the right support.

HINT: DO YOUR HOMEWORK

It's a good idea to meet the tech team halfway. Buy a "For Dummies" book on databases or read a TechCrunch article about Ruby on Rails. While business leaders should not pretend to know more than they do, it is worthwhile to show technology partners that you are trying to understand the matter at hand.

All of this is tough work. Trying to master reams of new information may make you feel like you're back in third grade, floundering with long division. I have found that many adults feel very uncomfortable when they do not have a sense of competency. It's a very vulnerable feeling—one that you may not have actually had since third grade. It can bring out the worst in people.

Have a Trusted Technology Partner

For the nontechnologist business leader, it can be critical to ally with a trusted technology partner. Business leaders who do not have a background in technology are well advised to partner with a senior technology professional; this may be your existing CIO or a technology consultant.

You might well ask, "Why do I need a technology partner? Don't I have all these techies around?" Yes, you do have a lot of techies; that is part of the problem. Remember, a few paragraphs ago, we left those techies driving all over the road. "A lot of techies" are not the same as a senior technology partner who can educate you and who has your back.

It's a good idea to think of choosing this technology partner as you would a specialist. Back to the operating room: Say you were facing some kind of complex surgery, like brain surgery. How would you select a neurosurgeon? You would talk to friends, do research, and ultimately identify a candidate. You would look at the number of surgeries performed and the rate of success. You would probably also be reminded that sometimes the best surgeons don't have the best stats because they perform the most risky operations. Then, you would look in the whites of that person's eyes and determine if you trust him or her.

Choosing a trusted technology advisor is virtually the same process. That person will talk business leaders through the complicated issues, provide high-level advice, and collaborate with the business on a strategy to accomplish your technology project.

How Best (and Worst) to Work with Your Technology Partner

You may have a relative who likes to argue with his doctor all the time. He brings every Internet study he can find online to challenge the doctor's suggested treatment. You may also have heard your relative discuss four other people he's contacted and what their doctors said. If you've seen this, you may be aware this is not the best way to work with a doctor.

A good professional appreciates your research and loves intelligent questions. He or she will answer them happily, taking no offense but instead, enjoying the fact that you are taking an active role in your case. However, when it gets to the stage of arguing or questioning *every* recommendation, things have gone too far; because in addition to all the data, the expert has hands-on experience in real situations that no amount of research and conversation can duplicate. Bottom line: Don't partner with an expert if all you are going to do is argue with them. You hired this person for a reason. Unless you yourself are going to get a degree in neurosurgery, you need his information. If you don't trust him, fire him.

Too Many Cooks

Getting lots of opinions is another strategy business leaders employ when they are confronted with a technology project. This seems very logical from a human nature point of view. If you don't feel confident in technology, what could be more natural than to bring more brains to the situation, and (perhaps subconsciously) share the responsibility for the decisions? Many people in organizations represent themselves as technologically knowledgeable. Why not harness all their opinions?

The constant gathering of opinions and attempts to build consensus simply do not work in some endeavors. Technology is one of them. Business leadership must rely on a limited set of voices—for example, that of a trusted technology partner—and decide.

It is perfectly reasonable to ask for opinions on different topics. For example, you may ask your technology partner to collaborate with an infrastructure expert with whom you have a relationship. Good professionals welcome such an approach, as long as distinct boundaries are set. You must establish who has the final say on what.

The business and its technology advisor need to have one clear voice and a clear direction, and this needs to be communicated to the teams underneath.

CHAPTER 5

Project Research and Technology Choice; Conflicts at the Start of Projects; Four Additional Project Delays; Initial Pitfalls

The information in the previous chapters has brought us to the point where we are truly at the project's starting line. Decisions have not yet been made, but they are about to be! For example, your company may know what it wants to do (replace an aging database), but how you want to do it or with whom must be nailed down.

Right now, at the starting gate, it's time to address many of the key issues introduced in Chapters 1 to 4:

- What's your project approach? Custom or packaged software? Phased rollout or single launch? Pilot or no? A single tool for many functions or different tools for each?
- Who is your team? What roles do you have in place? Which are missing? Will you go with in-house or external resources?
- Do you have a trusted technology partner on board to help inform business decisions?

Choice of Technology, a Definition

It's rare that a company says something like, "The purpose of this initiative is to install the Microsoft Dynamics accounting system." What people say is, "We need a new accounting system. The purpose of this initiative is to install the one that will best support our core business goals [list goals here]."

Maybe the correct accounting package is, indeed, Microsoft Dynamics software. Maybe it's something else. In either case, deciding on the project's main technology is called the "choice of core technology."

In some situations, additional technologies must be chosen as well. In the previous example, the company may also need a piece of reporting software. Or it may need invoicing software to feed the accounting system, as might be true in media companies that generate "insertion orders" through specialized software. In any of these cases, one technology is usually the main focus. Companies usually don't allow a preference on reporting software or some other tangential technology to determine their accounting system.

Choosing technology usually means researching the different possible solutions in the marketplace. It also means understanding your own business needs.

The Project's Research Phase

It is often critically necessary to do research before you can decide on core technology. Internal resources can be dedicated to doing the research, or an external expert can be hired. Either way, the research must cover certain topics:

- Current state
- Business need
- Possible solutions

Current State

This is a deep dive into the company's situation. Why has the technology need arisen? How is the current technology falling short? What is in place now and how is it operating? This part of the research is expressed as "understanding current state."

You may say, "Who cares about the current state? We're on to greener pastures!"

Not so fast. A business's current state inevitably has technological realities that must be considered. It often reveals a lot of "how we got here" information. There may be complicated reasons the current state technology is making everyone miserable. Often in current state research you find things aren't as easy to change as they might look on the surface.

Further, in current state research, you must look at the whole picture, not just the piece you are adding or changing. Usually the different technologies in an organization must fit together like puzzle pieces. If you are going to swap one out or replace it with a new one, you must understand the entire puzzle.

A good example is a media company with both print (e.g., traditional magazine) and digital (e.g., website) properties. Let's say such a company wants to redo its website. But the website feeds from the same information as the traditional print magazine. The team must know about what's going on with the traditional print technologies, and how it all works, in order to redo the website.

Often researching the current state involves digging up old documents and contracts and talking to that one guy who built the custom database. You must know what exists before you start to change things.

Integrations and Current State

I've used the word "integration" before. This is a place where two systems exchange data. If you are replacing a system, you also have to replace the integration point. If you don't, a business process will fail. In the real world, what this can mean is that credit cards don't get processed, invoices don't get out, or vendors don't get paid. If you have current state integrations, they must all be documented before you begin.

Data and Current State

All companies have data. They have thousands, usually tens of thousands, of Word documents and Excel files, customer records, and contracts. If you are upgrading systems, switching systems, or developing new software, you must consider the impact on your existing data.

Maybe you are rolling out workplace collaboration software such as SharePoint. All your documents will migrate from the file system onto the SharePoint platform. How many of them are there? What will be the new organizational structure? Are they all going to be brought over or are you going to "clean closets"? You must understand the current state of the data to unearth potential problems and make key decisions.

Here are some common questions usually asked about the current state of a company's data:

- Will the new system be able to read the old files?
- Will the old files come into the new system in a way that I can find them?
- Is the data "dirty," affecting its ability to easily flow or "migrate" into the new system?

The research phase investigates the current state of the company's data and provides information about how that data may be affected by the software project.

Business Needs

As the old saying goes, we're not in this for our health. We're doing a software project because of some business need. Perhaps even a business imperative.

At the inception of research, the business need may only be expressed in the most high-level terms. For example, the COO might be saying, "Our data storage needs are exploding, and we have to find a solution to deal with it."

What does it mean that the data storage needs are exploding? Exploding for whom? What affect is that having on business's ability to function? One high-level statement such as this often means dozens of examples throughout the organization, each with a slightly different nuance.

It's the job of the research phase to investigate and document these dozens of situations.

Hint: It's not likely every problem will be solved when the project is done. But the majority should be solved for the project to be considered a success. How can you know this if the full situation is not researched and documented?

Possible Technology Solutions

There may be dozens, if not hundreds, of possible technology solutions to address a given business need. Often large consulting firms such as Gartner Group will publish reports on what are currently considered the best technologies to address different needs. Such reports rank technologies such as digital asset management systems, content management systems, data visualization systems, and so on. This information inevitably includes considerations we've touched on: off-the-shelf vs. custom and one-tool-for-the-job vs. separate pieces of software.

Such reports, though sometimes pricey, can be helpful in the research phase. A CEO or board of directors may require you to know what the "heavy hitters" such as Gartner or McKinsey have to say on the technology terrain your business is considering. I have found these reports are particularly appropriate in the larger business space. For medium-sized businesses, you may get as much or more value by doing "journalistic-style" research, described later.

However you go about finding the information, the goal is the same: You want to line up the candidates and narrow them down.

Take the example of a content management system (CMS). A company has hundreds of choices. Nevertheless, there are certain main contenders in certain vertical industries. You must identify the candidates you want to evaluate and compare.

Demos

The research team can ask for product demos and specifications from the providers of the technology. Say you are investigating CMS systems, and you've identified one called Sitefinity. You can call up a programming group known for Sitefinity development and ask them for a demonstration of the products' features.

If your next candidate is Drupal, you would call up a Drupal shop and ask them for the same. Slowly you will build a list of information including features, costs, advantages, and disadvantages. The next step is a comparison grid.

Comparison Grids

A comparison grid is simply a table where the information about the different technology choices under consideration may be compared. You will map the technologies' features against the business needs and other considerations such as cost.

The first step in creating a comparison grid is, of course, to decide what it is you are comparing. What features and functions are most important for your company? What's your budget? What are your users like? Are they experts in technology, or do they struggle with complex products?

As these questions reveal, you are not just looking at business needs and product features. In point of fact, a product may meet 100 percent of your business needs, but take two years to roll out and cost $2 million. Despite its perfection, it's off the table. Another example might be a technology that addresses all business needs but that is very complex. If the people in your organization can't use it, you don't want it.

Table 5.1

Criteria	Product 1	Product 2	Product 3
Product features	***	**	****
Ease of use for business	**	***	****
Licensing cost	***	**	****
Cost of implementation	*	**	****
Resource availability	*	*	****
Ongoing cost	****	**	***

In looking at cost considerations, remember to consider both licensing fees and implementation fees. Some products may be "free" to license as in the case of open source technology, but be costly to implement. Another factor affecting cost is the availability of resources. If a technology is rare, it can be hard (and costly) to find resources to implement it. Finally, some technologies have expensive upgrades and other maintenance costs that must be considered. These future-related costs are often called "total cost of ownership" because they address the initial costs to implement plus ongoing costs to maintain.

Therefore, a complete list for a comparison grid (Table 5.1) should include the following:

- Product features compared to business needs
- Product ease of use
- Licensing cost
- Implementation cost
- Availability (and cost) of implementation resources
- Ongoing costs, sometimes called "total cost of ownership" (TCO) over a period of three to five years

Your own comparison grid will likely contain other topics important to your business. Table 5.1 is a starting point that all comparison grids should have.

Talk to Other People, a Journalistic Exercise

Talking to other companies who have faced similar situations as yours is tremendously valuable. In fact, some of the comparison grid information outlined above may be impossible to get without it. I have found that people are incredibly willing to talk and incredibly open with information. Of course, you must usually select people in a noncompetitive industry.

It's essentially a journalistic exercise. At conferences, on blogs, or in trade publications you can discover who has recently undertaken a project similar to yours. Sometimes the technology vendors will give you people to call. Then you call them up and ask:

- What was your business need? Why did you undertake the project?
- Can you tell me something about your "current state"? What was the technology situation you were addressing?
- What made you decide on the technology you picked? Do you have any regrets? What would you change?
- What things did you discover that you didn't know? What product features weren't as advertised?
- Did you go with an internal team or external? Why? If external, would you recommend them?
- What were your packaged vs. custom considerations? Why did you choose the [packaged or custom] route?
- Did you phase or have one launch? Did you pilot?
- What are your ongoing costs?

If you have any other assumptions to validate, such as whether the product is easy or hard to integrate with other systems or easy or hard for users, throw them in here. If you're lucky, the person you're talking to may tell you exactly how much the entire effort cost them.

As you can see from the previous list of questions, you are assembling information on all the topics covered in the first four chapters of this book. You are getting another company's most recent experience. Recent experience is key and not to be underestimated. Remember Moore's Law? We're back to it again. Usually reports and articles, even from illustrious consulting companies, have trouble keeping pace with the inexorability of Moore's Law. In technology, calling up someone who's done something exactly like or pretty close to what you are attempting nets the most valuable and current information available. The journalistic exercise is also helpful because you can speak to people in companies that are closely related to yours in size and business need.

Hint: It's usually best to do this kind of journalistic research when you have no more than four top technology-choice candidates.

How Do You Know When Your Research Is Done?

This is a very common question, though I've found people are often afraid to ask it. Don't be afraid! There's a simple answer: Your research is done when it starts to repeat itself.

For example, you might be looking at a few different sources to make sure you understand the features of a product. When your sources start to say the same thing and you are not uncovering much new information, you are done.

Or, you might be doing the journalistic exercise described earlier, and you've talked to two different people about their experience with a technology choice. On the third, you're hearing a lot of the same things the first two said. You are probably done.

Research Reality Check

Many companies labor under a mistaken impression: Once a thorough research and discovery phase is complete, the technology decision is all but made. After all, if you're trying to decide on a washing machine and you look up features and functions in *Consumer Reports*, it quickly becomes apparent which is the best fit for you. But when it comes to technology choices, the comparison grids often seem only to fuel the debate. We'll talk about "religious wars" later in this chapter.

Usually, a thorough research phase raises lots of questions. It presents you with a set of variables to work with, not definitive answers. This software solution is cheapest, but hardest to integrate with our backend system. Or this software solution has the best features for our business, but resources to implement it are scarce and/or expensive. These are the types of findings research often reveals. You pull one lever down, and another lever goes up.

You Can't Run the Control

If you were a scientist testing out a drug, you would have a control group. For example, you might want to answer the question, "Does this antibiotic work better than the standard treatment to cure infection?" In this case, the drug trial team would give the new antibiotic to one group of people afflicted with the particular infection, and the standard treatment to another group. The standard treatment group is the control. By comparing your new treatment results against them, you know if you have a better drug.

You can't do this in software development. In other words, you can't roll out your new website in Drupal and also in WordPress and see which effort took less time and less money. You have to go on the best information available. There are usually no definitive answers.

Hint: At the end of a software project, a group of people within the business will usually assert, "It would have been better if we went with [the other solution]." There is no way to avoid this, and there is also no way to know definitively who is right.

Religious Wars

I compare choice of technology to "religious wars," or, alternatively, to "red state versus blue state" conflicts. People are passionate. Positions harden quickly. Religious wars over technology are ridiculously common, affecting 90 percent of all the software assignments I have participated in during my career. You can set your watch by it.

The symptoms will often set in before the research phase has even started, as it becomes evident that some technology solution is favored by a key constituent in the business. Often this is driven by a bias of a high-level technology professional, such as the CIO, or a business leader, such as the CEO. But strong technology bias can come from another source such as a valued programmer. If Jane the programmer has been driving all over the road, given sway in technology decisions as the business stakeholders shirk, her preference for custom-coded Java will be voiced loud and strong. The business is used to deferring to Jane. Jane expects them to defer to her now.

In religious war situations, discussions on technology degenerate quickly, and the meeting room becomes a battle zone. Getting through this stage can take weeks or months, as business stakeholders become increasingly confused and technologists hurl jargon at one another.

There are many reasons for this situation, but one that is too commonly overlooked is pure self-interest. For example, there may be a vendor in the room who's worked with the company on other unrelated projects. Because the vendor knows the company, they are invited into discovery and research discussions.

But that vendor specializes in .NET. If the conversation drifts to consideration of open source technology that might be used in place of .NET, the vendor will aggressively defend the .NET solution. Not, perhaps, because it's the right tool for the job, but because his role in the new project is threatened. They know the company! They've done five projects. They know what's best in this project! Similar dynamics can occur with on-staff technologists who may fear they can't master a technology under consideration or that they might be replaced. As the discussion veers back and forth from open source solutions to .NET, an engineer might be sitting there thinking, "I have a job. No, wait, I don't have a job. Okay, now I have a job again."

I have found that business stakeholders are frequently blind to these job-security issues—often because, as noted previously, they mistakenly believe that all technologists can program in any language and master any technology at the drop of a hat. It's as if business stakeholders, at some level, believe technologists are like the robot C-3PO in *Star Wars*, who can listen to a few phrases of some alien language and instantaneously speak fluently in that language. This misunderstanding can lead to trouble. Businesspeople may continue to weigh technology choices and be completely oblivious to the various anxieties of the people in the room around them.

Passion over Reason

There is another factor that raises the temperature of technology-choice debates. Technology seems to inspire a passion in people that other topics don't. It is a subject for another book to answer the question: Why do technology projects attract huge amounts of passion, ego, and debate? But it is nevertheless true that they do. And I encourage any psychologists or sociologists who are reading this book to answer it. I don't know the *why*. But I do know the *what*. You need only walk into a dinner party and say, "Mac versus PC" or "iPhone versus Android," and by the end of the night, people won't be speaking to one another.

Similarly, a programmer who uses open source technology may view open source not only as a tool, but as a point of personal identity. "I am tall, blonde, a Democrat, and a believer in open source software." Since he comes from this perspective, he may not be able to have a productive conversation with a person who feels that the non-open source .NET programming framework is part of *her* personal identity.

Business Stakeholders and Controlling Ego

In the previous chapter, we discussed the need for business stakeholders to "step up" and become knowledgeable about the software project. This is necessary because the *business* needs to take ownership of its technology strategy and decisions.

We also noted that business leadership may not be very technologically oriented at all. This can be unfortunate, but it is nevertheless a reality. Business stakeholders must come up to speed, so they can make top-level calls such as technology choice.

But a common human reaction to feeling inexpert is insecurity. And a common reaction to insecurity is to try to prove you *really do* know what you're talking about. Regrettably, many businesspeople endeavor to show they know as much as the shamans—or even more! They already know everything about HTML5 and mobile compatibility. They read one, two, three, or seven books on it. This kind of reaction contributes to the creation of a religious war.

Business stakeholders add to the battle of egos with a techie on one side, spewing jargon, and a businessperson on the other, trying to prove his tech cred. Once you've reached this point, it's hard to achieve sound decisions, much less good outcomes. Business stakeholders can help break the cycle by reminding themselves it's okay not to know everything, and resist the urge to prove their smarts.

How to Stop a Technology Religious War

It's critically important for business and project leadership to be able to recognize a "religious" debate over technology. The information already given in this chapter helps with that. The earlier descriptions should help you recognize if a religious war is afoot.

But how do you stop it?

We'll discuss how right now.

Not So Easy

I won't kid you. It's very difficult to stop a technological religious war once it has taken hold. It's much better to avert the conflict in the first place. As with many things, prevention and early detection is the cure.

Preventing a Technology Religious War

Call it out. It's as simple as that. When I partner with a company, I point out before we even begin that technology generates passion and a "religious war" is likely if not inevitable. This kind of up-front conversation serves as an inoculation. Then you can discuss tools to handle the situation and the realities, sometimes unpleasant, that everyone must face.

Here's one of my adages: "Passion is for amateurs." Enthusiasm for a particular technology is great ... up to a point. But when it goes beyond a keen recommendation to the ardent, avid, passionate, or obsessed? Beware! Seasoned software professionals know that technologies come and go, and that many different tools can be used to accomplish the same job. Therefore, passion is for amateurs.

I frown on overly passionate technology recommendations. I state the "passion is for amateurs" rule up front. Set your boundaries at the beginning: The business will not tolerate red state/blue state behavior in discussions regarding technology choice. This one rule will go a long way to heading off the religious wars and getting through what can be a long initial delay over technology.

Make it clear the business leadership and key business stakeholders will make the call *and* take responsibility for the call. Shouting jargon is not helping them make their decision.

Being Right

A wise person once said, "You can be right, or you can be married. Pick one."

This saying is as true in business relationships as it is in personal ones. Everyone on the project team is getting into a long-term relationship with everyone else and with the broader business. It's of little consequence if a faction prevails in the belief they are "right" if relationships are so damaged afterwards that no one is speaking to one another.

As noted earlier, in software choices, you cannot run a control and prove who was right anyway. There are likely many tools to accomplish the same job. All will involve compromises. None will eliminate pain.

At the outset of a project, it's helpful to note to the team that many discussions will come up and many times people will think they are "right." Ask teams to think of "you can be right or you can be married" when they are *sure* they are right. Most of the time, they will grin sheepishly as if remembering the way they behaved in a conflict with their spouse over the weekend.

Stopping a War in Its Tracks

Business leaders, program managers, and project managers who spot religious wars have to pull the emergency brake and find a way out. Project leadership must flag that the conflict has gotten out of control and map a way out of the morass.

The project leadership must deal with the people in the room who are simply defending their own self-interest. You must recognize who is merely trying to protect her or her own job. Then, take them aside and be honest with them. They need to know the score, such as, "Yes, it is likely if we decide for open source you will not be the vendor of choice. I invite you to participate in the discussions around core technology choice and tell us about the advantages of the .NET approach." Next, project leaders must make sure everyone is clear on his or her role in the technology debate.

In order to have a productive debate, everyone's position must be on the table with no hidden agendas. The most basic component of a debate is having voices to represent the different sides and to clearly know who is representing which side—in this case, the possible technology solutions.

In the previous example, the business with a .NET expert is being clear that his or her role is to give information on the .NET *solution*. Making sure that person is clear about his role of supporting the case for .NET also makes clear he is not to represent himself as an objective voice for the best technology. If he has worked in .NET all of his career, he cannot really speak to an open source solution. This kind of strategy is particularly needed when a company has a long-standing vendor partnership or an in-house programming resource who is used to being regarded as the "expert" on all things technological.

If you discover you need another voice in the room to support the alternate point of view, go find that resource.

Détente and Finally Ending a Technology Religious War

The best way to end a religious war is to declare a winner. As noted in this chapter, the *business* gets to make the decision on technology choice. The highly trained technologists may end up with their noses out of joint because people without computer science degrees are making a major technology decision. They will have to get over it because these are just the rules of the road. The business must own all strategic business decisions, and a choice of core technology is a strategic decision.

Clarity

Ideally the project leadership would make it clear from the beginning who gets to make the final call about choice of technology: the business. The rules of the road in this ideal situation are clear up front, and everyone knows their role in the debate.

However, we don't live in a perfect world, and I'm sure many people may be picking up this book in the midst of a religious war. Clarity is best up front, but it helps whenever you can introduce it.

To review:

- Programmers, analysts, and other technologists may *provide information* and *advocate for a solution*.
- Project team leadership digests and summarizes this information (for example, in a research document) for business stakeholders and leadership to consume. The project leadership team takes the various recommendations and provides a *summary recommendation* with all options outlined as in the comparison grid.
- Business stakeholders and business leadership *decide*.

The project leadership must be explicit about how the final choice will be made. Research will be read, and technological expertise will be considered.

Business stakeholders and leadership must also be clear that they are on the hook for the final decision and will take responsibility for it.

The Role of the CIO

The CIO may be the only one in the business leadership group with a computer science degree. Therefore, her voice will undoubtedly be a strong one in influencing the leadership group about whether the business is choosing Microsoft Dynamics or Sage, .NET or Java, Drupal or WordPress.

Sometimes, however, the CIO's voice is overly strong, and this can cause problems. For one thing business leadership and key stakeholders must *all* own the technology decision. As I've touched on before, it's a common problem for businesspeople to shrug when it comes to technology, avoiding decisions because they feel unqualified, therefore shirking their duty. An analogy would be allowing your accountant to make all your important financial decisions because you aren't good at math. Business-level technology decisions can't all be on the CIO.

Further, the CIO is a human being and also has her biases. There's nothing wrong with a bias, as long it's transparent to everyone.

Let's say the company is considering which e-commerce package to implement. The CIO is an expert in a software product called Magento. Her experience is deep. She's implemented it before in other companies with success. She knows the resources and the pitfalls. Everyone in the business likes the CIO, and she has an excellent track record. Using a technology in which she has expertise creates a situation where she can serve as both project sponsor and program manager of this effort. She will get it done.

At this point, if the Magento solution is anywhere within 10 percent to 20 percent of the cost-range variance of the other solutions, it might be a good idea to just throw the comparison grids out the window and make the call for Magento. The CIO's expertise and previous experience is invaluable and will likely save money in the long run. We'll touch more on the importance of the human element later.

On the other hand, some CIOs are what I call "empire builders." Empire builders are looking to increase the power of technology inside the organization. They may want to build an impressive new software solution to leverage themselves into their next job. The may have a preference for a technology because they know it's a hot ticket in the marketplace, not because it's the best solution for the business. In this case, biases are hidden and the influence of the CIO's voice in the decision is not beneficial to the business.

Two Most Important Factors in Core Technology Decisions

In my experience, the following two factors are the most important to consider in technology choice. The first is obvious, the second less so.

- Budget constraints
- Teams and the human element

Budget Constraints

You don't have a budget yet. Exact budgeting will come in Chapter 7. But your research should have revealed a high-level sense of what the budget is likely to be. If you have done the journalistic exercise of calling other companies, you may have a more concrete sense of the cost of one solution versus another.

Research frequently reveals the most cost-effective solution. There may be some feature compromises, but those are often acceptable trade-offs if the next likely technology solution costs 50 percent more. No one is going to consider installing an accounting system whose annual licensing fee exceeds the company's annual revenue. Simply put, your budget will define the world of technological choices you have and often make the final choice.

The Team

The team of professionals who will be implementing the solution is, without question, the single most important factor in the success—or failure—of your technology project.

That's why I have discussed team and roles and personalities in such detail in Chapter 4. Frankly put, if you have a good software team with good account management, good planning, and excellent communication, the game is almost already won. Though you need *all* these elements in place, programmers are the most foundational. No amount of project management and communication can overcome bad code.

As noted when we discussed programmers in Chapter 4, it's hard to know if you have bad coders or implementers. Sometimes the only way to know is to look at track record. Has the programming team accomplished a similar task successfully?

In choosing technology, therefore, the best strategy is to combine the technological with the human. In other words, in the projects I work on, I intentionally allow the skills and preferences of the *best available team* to influence—and make—the technology choice. When it comes to technology, all senior technology professionals know that there is almost never a single "right answer." Many technology paths can lead to the same route, and none is without its pain and compromise.

Nothing matches comparable experience and a proven track record for reducing project pain and achieving project success. Literally nothing.

As we will discuss in future chapters, in the highly complex world of software development, things *will* go wrong. Many things will happen that are unexpected and entirely out of your control. The *one and only* hedge against this is a team of proven, reliable, and experienced professionals who have been there before.

Choosing Technology and What *Not* to Consider: The Future

Yogi Berra once said, "It's hard to make predictions, especially about the future." Nothing could be truer when it comes to technology.

In the section about comparison grids, I included the concept of total cost of ownership (TCO). To review, it's an estimate of what a piece of technology may cost in the future.

However, total cost of ownership must be taken with a grain of salt: in fact, many grains of salt. Remember Moore's Law? Things are changing and fast. People implementing all kinds of technologies in 2006 had no idea the iPhone was about to hit or how big it would be, changing the technology landscape, creating all kinds of needs, and eliminating others. Ditto for the rise of the tablet. Whole technology strategies and implementation plans were upended by the rapid adoption of these two platforms.

Spending too much time on future considerations or giving too much weight to total cost of ownership is a mistake many companies make, sometimes allowing it to be the major influencing factor in technology choice.

In addition to TCO considerations, people can spend huge amounts of time on topics such as "scalability" or "extensibility." You will frequently see those particular words used. The implication is that it's not just a question of what will the technology do for me today, but also what will it do for me tomorrow when I grow big and have greater needs.

I have found when companies get wrapped up in buying a technology for the future, it leads to trouble and wasted money. Sometimes I feel like I'm breaking some unwritten business rule by saying this. But the truth is, technology is changing so fast, it's almost impossible to predict what tools and options will be available a year or two years from now. Most software development projects will take six months to a year to complete. Therefore, in the realm of software development, it actually makes *more* sense to think in the short term.

Other Conflicts that Delay the Start of Projects

You may be starting to feel the hardest part of the project is just getting out of the starting gate! If you're feeling this way, you're right. It can be very hard. Multiple phenomena, in addition to technology choice, may delay the project's start. It's important to grapple with them because if they are not resolved, they can have the effect of "poisoning the project from within." There will be more about that in Chapter 12.

By far the largest initial project delay is technology choice. But there are four others to be aware of:

1. Business strategy and organizational authority
2. Blue sky
3. Overanalysis
4. Design

Business Strategy and Organizational Authority

Sometimes company strategy is not clear. Or there may be conflicts among the business leadership about what the company strategy is. Other times unofficial fiefdoms exist within companies; authority structures and reporting lines are blurred. They may be clear on the org chart, but in practice? Forget it.

Medium- to large-size technology projects often surface these issues. Following are some examples.

We've talked about implementing CRM solutions. To review, a CRM solution holds all of a company's contact data, from employees to customers and everything in-between. Let's say ACME Inc. has decided to implement a new CRM. The group responsible for consumer sales is ecstatic. But the group responsible for B2B sales (business-to-business) is upset. They have their own CRM tool that's working just fine for them. What do you mean they have to give up their tool and control of their data? They know the consumer group is more powerful and better funded. How will their B2B needs be taken into consideration? With every feature request, they will clearly lose to the consumer group!

Another example: A company has policies in place so that employees get standard laptops or desktops depending on their job function. A vocal division in a company has won a concession to have different machines, for example, Macintoshes. A business system is being rolled out that won't work well on Macs, requiring plugins and compromises. This unit is now up in arms. Will they have to give up their machines? Their division head says, "No way. The business must find a different system."

Another example: A magazine has a subscription model requiring customers to subscribe to and receive the print edition in order to receive access to the online edition through an unlock code. The subscription software is old and is being re-implemented. The powerful editor in chief insists it must continue as it always has. The project team discovers the costs are much higher, since most digital-and-print publications allow for a digital-only subscription. It will necessitate special coding and a new process with the fulfillment vendor. The editor in chief is incensed.

These are all business strategy questions. They are also governance questions. In many cases technology will cause the interests of different divisions to collide, and someone higher up on the organizational chart must make a decision. If business units have been operating autonomously, mostly allowed to "do their own thing" as long as the numbers are met, there can be trouble.

Hardly ever does a project get off the ground without some tough decisions and core issues surfacing at the beginning. Many times, software development projects are a lightning rod for conflicts that have more to do with these kinds of overall organizational authority questions or strategy than they do with the technology itself.

It's important to face these conflicts. Striving for consensus is admirable, but the absolute worst thing you can do is gloss over whatever fundamental debates surface at this time. Papering over strategic disagreements will come back to bite you later. In my 20 years of experience, I have learned that the conflicts of this nature surfacing early on are almost always mission-critical and must get solved. If they don't, the lack of key decisions can doom the entire project.

You may confront issues and conflicts masquerading as other, more benign things like the design of a menu bar. But core strategy can be at play. For example, an argument over the menu design may be masking some serious divisions among organizational silos, or differing viewpoints about who the company's customers are and how to best reach them.

HINT: TEMPER, TEMPER

If the tempers are getting hot, you're dealing with a core issue that needs to get resolved. When you notice these flare-ups, it may be time to set aside the menu-bar design argument and address the proverbial elephant in the room.

Design

Unless your project is purely "back end," such as a database implementation that will never be seen by anyone outside the company, there will be some design involved. Design is a huge sticking point and can cause delays for weeks and months.

One of the reasons design is such a logjam is that business stakeholders tend to feel they understand design and usually have experience with it—in contrast to the difficult technology discussions where they may feel at sea. Thus, many people gravitate to discussions about design as a kind of safe harbor.

Furthermore, many business stakeholders mistakenly believe if they just know what the software or website will *look* like and get that right, the project is all but done. So they spend weeks working with designers who may or may not understand the way the visuals must interact with the technology. Teams generate sample screens that look "just so," only to find that they cannot be implemented technologically. Precious time has been wasted on the largely superficial activity of design, and the programmers have yet to get started.

In an upcoming chapter, I will further discuss strategies for integrating visual design into your project. It's important to remember that software is interactive, with the user's delight in the end product coming principally in how it clicks, moves, and behaves. In other words, design *enhances* the dynamic software experience. In general, you want to avoid over-focus on design in the early stages, and get onto other kinds of visualization as described in Chapter 6.

Blue Sky

"Blue sky" is characterized by teams of people talking endlessly about possibilities—what incredible things the software could do that there will probably never be time and budget to do. There is a big danger in "blue sky-ing" beyond the obvious waste of time. As we will see in upcoming chapters, one of the most important imperatives to getting through a technology project is controlling scope. Do something modest, get it right, *then* do more. But somehow—I think because technology inspires excitement and passion—people want a big impressive rollout. They want neat and cool. They long for the "killer app."

Many technology projects aren't meant to be killer apps. They are what we call "re-platforming." This means an existing piece of software must be reprogrammed because it will no longer be able to run—for example, if it was originally programmed to run on an early version of Windows. Or perhaps an existing process needs to be web-enabled. Too frequently, folks will take this opportunity to reinvent the software and pile things on top of it. They want to know *what more* it can do.

Blue skying must be nipped in the bud. After one or two brainstorming sessions—a reasonable number—bring people back to reality by reacquainting them with the likely calendar and the budget, still rough at the beginning phases. Remind them that most people are lucky to successfully complete a *modest* software project, never mind a "blue sky" one.

Overanalysis

Overanalysis is less common than design delays or blue sky, but it can still be an obstacle to success. Most organizations are too light on analysis and discovery and would benefit from lots more documentation—in particular of the visual kind, as we will discuss in Chapter 6.

But while the "analysis paralysis" myth is overplayed, it does happen. Some organizations have a culture that promotes it. Overeager business analysts feel that the way to decrease risk is to discuss and document every detail down to the atomic level. Sometimes there is a decision maker who just can't pull the trigger. She will send teams into layers and layers of analysis.

A good rule of thumb is this: If there is so much analysis and documentation that a reasonable person could never read it, there is too much.

The Project Charter, a Key Document

After all the research has been done, conflicts and barriers overcome, and religious wars settled, it's time to create the project charter document. This is the highest-level document of the project. It says what you're doing, why, and how.

Here are the elements of the project charter document:

- Name of the project
- Purpose of the project and business case
- Project team, including project sponsor, key business stakeholders, and all other team members covered in Chapter 4
- Project approach, including technologies chosen and why
- Project goal date and high-level milestones
- Cost assumptions based on research

Many companies skip right over the project charter document, moving fluidly into scoping. I've found it can be helpful to formalize things at this juncture. After all the research and debates, it's important to say who will be doing what and why. Often, the most important purpose of the project charter is for the business leaders to authorize proceeding forward. In other words, they have had a look at likely costs (not a budget yet!) and goal dates (not a timeline yet!) and have given the nod to move forward.

Conversely, if the goal date and research-based budget are way out of line with business expectations, getting a strong "No!" at this point is critical. It may be disappointing to the project team, but it's much better than going into all the work involved in scoping (coming up next), only to discover the cost of the project is out of line with business expectations.

CHAPTER 6

Final Discovery; Project Definition, Scope, and Documentation

Y ou have reached the phase of your project where it is necessary to pin down scope, timelines, final project plans, and budgets and document it all.

You may say, "Wait! Don't I have a ton of documentation? I went through a whole research phase!"

Yes you do have a lot of documentation, and good for you! Right now, if you've followed the advice in this book, you probably have material documenting your "current state," the technology you already have, its quirks and considerations. You also have investigated your choices of technology solutions. You've got journalistic material, researched from the field, to help you decide. You may have talked to vendors and other resources needed in your project. You probably even have high-level budget numbers about what it will all cost. You may have a project charter or "vision" document. But you do not yet have a plan. We're about to get you one!

This phase of getting to the final plan is often called "final discovery" or "determining final business requirements." I have found people find these terms confusing because a lot of investigation and discovery has already taken place. It can also become a sticking point, especially if a vendor has been engaged in this activity. A client is asked to invest more in "final discovery." The client may say, understandably, "*Final* discovery? What have you been discovering all along?"

Final discovery simply means the last phase necessary to get to the detailed plan of action.

Budgeting and Ongoing Discovery; Discovery Work Is Real Work

A basic conflict in budgeting for software development is the fact that you need to make an investment to even *get* to a budget. Business analysts and mock-up people need to be engaged. Programmers need to be consulted, pouring through lists of specifications, thinking about how long various features might take them to code, and providing estimates. A project manager is collecting this information and sifting through it all, trying to make sure nothing is forgotten (e.g., "Wait, did we estimate the QA effort?"). Finally, when the dust settles after all this activity, you will have a solid requirements document and a budget.

In order for most businesses to be successful at software development, a shift in thinking needs to occur—probably by you, but likely by your management as well.

A common perception in both the research phase and final discovery is that "nothing is being done." You emerge from this phase with a document and a budget, but really that is "nothing." The "real work," in most people's view, has yet to occur.

Here's the shift: The discovery work *is* real work. It' the first third of a software project: I mean this in every sense—time, money, and actual work. Moreover, if you emerge from this phase with a clear, detailed business requirements document, you will save money in the long run.

You can prove it to yourself. Take your documentation to a software company, and two results will always occur:

- The software company will be able to budget accurately for your project.
- The number they come up with will be 30 to 50 percent less than if you started from scratch. Just ask them how much it would have cost for *them* to do the documentation and discovery.

In other words, you will pay for this work one way or another. There is no way to avoid it.

Budgeting Final Discovery

There are really only two tactics that work here:

- Use your own staff to do the final discovery.
- Pay someone else to do the final discovery.

Using your own staff is obviously the cost-conscious way to go about this. Ask yourself, is there someone on the team who can do the business analysis, talk to the business stakeholders, and find out their needs? Do I have some visually skilled person who has a basic understanding of software who can do lots of visuals (more on this later)? Can that person also communicate requirements to a technical team?

If you can staff it this way, you are fortunate. You will end up with thorough final requirements. If you are not able to in-house staff this work, you will need to hire someone to do it. That person or team need not be the ones who will develop the software. The assignment will be very specific: Come up with the business requirements documentation so that the project can be accurately estimated in time and cost. A good technology consultant can do this.

What Discovery Costs

The good news about initial and final discovery is that the costs are pretty predictable. In all the software projects I have been involved in, discovery is almost always one to three months. This is directly related to the size of a project. A project in the $5 million to $20 million range can have six months to a year of discovery. A project in the $500,000 to $1 million range usually needs about three months of discovery. A project of $250,000 to $500,000 can most times accomplish its discovery in six to eight weeks. A project of less than $250,000 can take a month or less.

What Comes Out of Final Discovery: A Plan

Your plan can also be described as your "scope document" or "final business requirements document." If you are working with a vendor, they may call this the "scope of work" (SOW).

All these terms and phrases are just another way of saying "the plan" that answers the following questions:

- What specifically are we going to do?
- What features and functions will the system have?
- What features and functions will be left out?
- Who is doing what?
- What are the dependencies (interrelated parts) we must consider? And what order must they go in because one thing must be done before another can start?
- How does all of this fit into a timeline?
- What are the unknowns and risks?
- What is the budget for all of this?

Note: We will deal with budgeting in Chapter 7. This chapter sets you up to create a budget.

Getting to a Plan

Getting to the detailed plan can feel tough. Here's why.

I have found that human beings are pretty good at the high level, the 30,000-foot view. The path we've walked so far, you may be surprised to know, only gets us to that high-level view. Yes, with all the conflicts and negotiations and delays, where you are right now is still at the high level. You may know, for example, you are replacing your current accounting system. Your research has revealed the hitches and hiccups you are likely to deal with. Still, though, from a software development standpoint, it's conceptual, not concrete.

Here's an analogy. Let's say it's 1850, and you've decided you are going to travel from New York to California. You have a good map. You know you will cross the Mississippi and the Rockies. You've bought yourself a good horse and covered wagon. But that's all. You have not identified the inns you will stay at, what supplies you need in your covered wagon, how you will get across the Mississippi, or the gear you will need to cross the mountains. You also don't have a plan B. What will happen if the weather turns against you? Will you need to take a different route?

This is the juncture at which we find ourselves. We have a map, but no plan.

The Murk

As mentioned earlier, human beings are comfortable at the high level, the map. They are also really comfortable when they have a detailed list. Even if that list is daunting, it's a list, which is solid and knowable. Literal project managers, in particular, *love* lists!

Human beings are *not at all* good at deriving the detailed list from the high level. I call this "being in the murk."

A classic symptom that you have entered into "the murk" will be heard from your literal project manager. She will be demanding the list. Taking the 1850 travel analogy, she will be asking, "How many grappling hooks and snow shoes are we going to need to cross the Rockies? Where are the canteens? Who's going to get them? Are these horse bridles sufficient or will we need a couple of spares?"

When she asks this, the rest of the project team, "the travelers," will look around stupefied. How are they supposed to know? They have the same questions she does! Everybody looks at everybody else expecting the answers to be provided.

What's happening here is the PM is driving to get the list of the "final requirements" before people are capable of giving them. She is avoiding the discomfort of the murk by trying to drive right from the high level into the lowest level possible, the list of tasks that must get done. She's leaping from one comfort zone (high level) into another comfort zone (lowest level). If you read the personality section on project managers (Chapter 4), you'll understand why. Literal project managers are often black-and-white thinkers. They are most comfortable when the tasks ahead of them are clear. Then they can sweep them up into tracking and reporting systems.

We're not ready for that yet.

Noticing the PM's discomfort is a good clue for the project leadership. It's a great flag letting you know you've entered the middle phase, the murk.

Simply put, the murk is that uncomfortable phase between the high level and the final business requirements. It feels unclear and unwieldy. People on the project team are not sure what the process might even be to get from where they are at the high level down to scope.

Getting Out of the Murk

I'll use the example of a recent project on which I was program manager. It was a rollout of an entirely new website, involving a new content management system. There was a data migration involved to move the content from the existing site to the new site. In addition, there was a major integration. The new site would need to talk with, "integrate with," an existing back-end database, managing all the company's customers and much of its finances. In some cases the new site would need to "ask" the back end if a user could have access to content. In other cases, the new site would need to take money and register people to become customers.

The project involved multiple teams. There were the experts on the back-end system. There were the experts on the integration part. There were the folks designing the new website. Everyone needed their own plan. To develop their own plans, each needed a detailed understanding of every other piece. No one had a final list. Everyone wanted the final list.

Classic murk.

In my role as program manager, I created what I call a "plan for the plan" document. It's not a project plan, but rather an overview of all the things that need to go into the project plan. In it, I took a guess at the timeline and also at the "dependencies"—what parts of the project were tied to what other parts.

Here's an example of the document—edited, of course, for confidential material.

The Plan for the Plan—Company A

Hosting

- Identify hosting partner
- Contract, licensing, negotiations
- Plan for integrations
- Test for performance
- Dependencies—multiple, especially integration and performance testing
- Timeline considerations:
  - July–Aug: Plan identify
  - Sept: Environment ready
  - Dec: Load testing performance testing

Content Entry

- What is needed over and above the CMS content entry interface?
- Workflow
  - Approvals
- Taxonomy and tagging
- Dependencies—content editing and entry, UAT
- Timeline:
  - July/Aug—Requirements
  - Sept/Oct—Build
  - November—UAT

Search

- Define desired search behavior against all targeted data sources
- Design faceted search interface
- Test against sample migrated content (dependency on taxonomy and content migration)
- Dependencies
  - Taxonomy
  - Content migration
  - Integration with backend (for permissions on content)

■ Timeline:
  • July: Pilot with sample content set
  • September: Build
  • Jan/Feb: UAT

Content Pages and Features

■ Review documentation of current state for all current features and
behavior
  • Identify stay/go/modify
■ Finalize feature set for new site
  • Define and specify all expected behavior and for which user types
■ Dependencies: Multiple
  • All UAT
  • Editorial interfaces
  • Content migration
■ Timeline:
  • Aug: Review/plan
  • Sept/Oct: Build
  • UAT: Nov

Integrations

■ Identify all integration points and expected behavior
■ Testing
■ Load testing and performance (hosting dependency)
■ Dependencies: multiple
  • UAT of any pages with integration features
  • Hosting environment
■ Timeline:
  • June: Pilot/design
  • Aug: Full business requirements
  • Sept: Build
  • Oct: Test

Back-end System

■ Review research to decide which features stay in the back-end system
and which move into the content management system
■ Dependencies: Front-end pages
■ Timeline:
  • July/Aug: Identify pages; refactoring necessary
  • Sept: Refactoring

- Oct: Reskin
- Nov: UAT

Data Migration

- Review/establish scope
- Create migration plan
- Dependencies
  - Integration testing
  - Taxonomy
  - Search
  - Editorial workflows
- Timeline
  - July: Migration #1 for pilot
  - Dec: Migration #2
  - Jan: Migration #3

If this looks like a brainstorm, it is! You may not even understand all the pieces, but that doesn't matter. It's a collection of all the things I could think of that needed to go into the project plan. It's not perfect. But it's a start. Working with the plan for the plan, you can get through the murk.

How Anyone Can Make a Plan for the Plan

In the previous example, since I had worked on many similar projects before, I was capable of producing the plan for the plan myself. In other cases, it's useful to assemble the team and brainstorm together.

The key questions are:

- What are all the components of this project you can possibly think of?
- What are the items involved in the part you just mentioned?
- Where is one part dependent on another?
- If you had to guess, when would each piece need to be completed to make our goal date?

Experience helps here, of course. But you can just have team members throw things out and capture it all in your notes. (There is more about good note taking in Chapter 9.) You may get uncomfortable reactions from your team members, programmers, PMs, BAs, and others who wail, "But there is so much! We can't possibly think of it all."

If you hear this, just keep asking, "What else? What else can you think of? Anything else?" Remind them it's okay not to think of everything. Encourage them to just free associate. You will hear things like, "Oh, yeah, we have to make sure we cut over the DNS. Wasn't Mary in charge of that?" It's a step. It's important. Write it down.

You might hear things like, "No one really knows how many documents we have or what shape they're in!"

To that the answer is, "Great. So there's a step on the project plan where we need to go in and analyze that." It's a step. It's important. Write it down.

Or you might hear, "I mean, I know there are supposed to be all these new features. And we've talked about it a lot in meetings. I can't remember them all. What are they anyway? Has the business decided on the final feature set?"

To that the answer is, "That's a great question. So there's a step to take out everything that's been proposed so far and present it in a document for a final decision." It's a step. It's important. Write it down.

Your literal PM will be feeling a lot better by now.

Different Approaches to Elicit the Plan for the Plan

The previous scenario serves as an example based on one of my projects. Your own situation may be different. Depending on the project, working through the murk may have its own nuances. Certainly, in my experience, this has been the case.

Another example: On a recent project, the high-level objective given to the project team was to integrate two systems with a third one. The company had just implemented a new CRM system, which was to be the "system of record," or master system, for all members in the organization.

The two other systems were order-taking and management systems. The high-level directive was for these two pieces of software to become "downstream" systems. In other words, they would no longer contain this data, but rather request data from the CRM for any person they needed to know about, such as a member of the organization or a customer.

In tech talk, this was a "pure integration project."

The teams knew the high-level picture: Get all data from the master CRM. But what was next? Everyone entered the murk. I heard things like, "This is nontrivial." And, "There will be a lot of downstream effects." And, "The receiving systems will have to be refactored."

Okay, but how is it nontrivial? What downstream effects? And what exactly will have to be refactored?

The PM started to pound her fist (metaphorically speaking, of course; she's a lovely woman), asking for the final list of what had to be done by each vendor. The vendors and other partners, however, wore that stupefied look and went back to "a lot of impacts" and "nontrivial."

What we did in this case was to provide screen grabs from each system and to display them to the entire working team. We showed an example of what existed in one system and in another. We started to make lists of where the source of data would change and what programming would be impacted.

Occasionally, during this process, someone would say, "But that's just an example! That's not all of it!" True enough. But, inevitably, if you follow this process, a picture will start to emerge that is much clearer than the high level. In the integration project, by using examples, we captured the major tracks or "big buckets" of the project.

You're looking, in other words, for the categories of work that need to go on and some idea of the nature of the work involved. You're not to the list yet, but rather the types of lists that need to get filled. Once you have these, you can move on to separate sessions where you really get into the detail on the major categories. These are the breakout sessions, covered in the next section.

The plan-for-the-plan phase may look different depending upon the situation, but you know it's time for the plan for the plan if the following things are true:

1. The high-level goal is defined. Significant research is in place.
2. A final list of business requirements does *not* exist, and the team is befuddled on how to get to it.

Getting to your plan for the plan always involves:

1. Brainstorming with knowledgeable team members
2. Getting it all down on paper
3. Identifying further breakout sessions

Exception to the Murk

Before we continue, I will take you down a quick detour. There are some software projects where no murk exists, and the list is known right off the bat.

These projects are typically upgrades or improvements to existing systems. For example, a group of salespeople might get together to discuss the company's business-to-business website with the technologists who run it. The salespeople have specific customer requests for how the site can be improved. A few potential customers may have even said to the sales team, "Hey, if you introduce this feature on your website, I'll sign right up as a customer."

If this is your situation, count yourself lucky and celebrate! The customer requests are your project scope. You have the list! Your delighted literal project manager will write it all down, and the technical team will estimate the effort (budget) and time (timeline). With business approval, you are off to the races.

For much of the time, though, on medium to large software projects, you aren't lucky enough to have the list from the get-go.

Breakout Sessions

Once you have a plan for the plan document, move on to the next step, the breakout sessions.

The Weeds Are Where the Flowers Grow

I have noticed a trend in business over the last decade or so. The phrase, "Let's not get into the weeds," has become more and more common. So much so that you might come away with the impression that any detailed discussion is a bad thing. It means you're not being "strategic."

I am impatient with this perspective, particularly when it comes to software development. I counter it by encouraging teams to consider an alternate phrase, "The weeds are where the flowers grow."

There is a time and a place for getting into the weeds, and it's right now, when you need to drive from the high-level concept of the project into specific project requirements. You must get into the weeds to get out of the murk. Breakout sessions are where you get into the weeds.

As you generate your plan for the plan, you will identify the major categories of work in your project, and almost all of them will need more discussion. You've been working largely with samples of the kinds of things that need to get done rather than the details and exact pinpointing of each item.

Breakout sessions are the multiple meetings on each of the main tracks in your plan for the plan document. Out of the breakout sessions will come the detail you need to make the final project plan.

Once again, you may sense resistance from your project team that they will never be able to define it all. To some extent that's true. Only in the most "waterfall of waterfall" efforts does the project team get to the fine detail on every task that must be completed. As you may sense by now, I am not an advocate of striving for this kind of fine grain.

I tell project teams (if you'll excuse the vulgarity) to "Barf it all up." List everything you can think of in the breakout sessions. Bit by bit, your picture will start to get detail and color. Some parts of the picture may become comprehensive. Some of it might still be a pencil sketch. That's normal.

Not All Questions Will Be Answered

By the end of the breakout sessions, you will have enough material to put together your project scope, timelines, and the other documents we're about to discuss. It may be you do not have all your questions answered. In fact, it's likely you won't. In order to get to avoid endless discovery and overanalysis, you're going to have to make some assumptions.

What assumptions? What is the right level of detail? These are common questions. The metaphors we discussed early on—piles of snow, Ikea desks, and heart surgeries—come in handy here.

Remember that Ikea desk, the one that came with all the parts and tricky directions? Boy, will you be glad to see those at this stage and so will the PM. In my experience, Ikea desks, the complicated parts of projects with many facets and a high degree of detail, present themselves readily in your breakout sessions. You can watch the rows of Ikea desks line up outside the meeting room door, each with its own set of weird directions and bags of parts.

Let's take an example of a breakout session regarding the integration project. The two existing systems ("legacy systems") had a bunch of contact data, people, places, offices, and hierarchies. They would no longer store all this data, but rather get it from a new system. During the breakout session on this topic, the team readily mapped out all these entities, decided on how the web services (ways the systems must talk to one another) would work, and identified how the two existing systems would change to receive the data from the new third system. It was complicated, but the detailed picture presented itself like a set of Ikea desk directions.

Piles of snow, you might be surprised to learn, are often less knowable than Ikea desks. It is common at this stage to have identified one or a few piles of snow. You know they are there, but you don't know how big they are. Nor do you know if it is fluffy snow and easy to shovel, or wet snow and heavy. Take the topic of data migration we've touched on multiple times. Often, data migration is a pile of snow exercise. After the breakout session, you may know you have to do a large migration of data from one system to another, and maybe from many sources. But you still don't know exactly how large the effort will be. How many pieces of data are there? Can the migration task be scripted, done automatically, or will it be done by hand? Is the data clean or dirty, difficult or easy to migrate?

If you have a pile of snow of indeterminate size, it's appropriate at the scoping stage to make some assumptions and move on, while noting, of course, you have made an assumption.

You may also have spotted a few heart surgeries. Sometimes, project teams will call these "black box" elements. You may know, for example, that you must work with an older or custom piece of technology, integrating with it or using it to develop something else. The piece of technology was developed in 1998 on Windows NT and has not been upgraded in years. The guy who built it is not even around anymore. No one really knows what will happen when they "lift the lid" on the "black box." This is another area where you will make "how hard" or "how long" assumptions and then note those assumptions in your scope document.

To review: Breakout sessions should unearth enough detail to create final business requirements and project scope. Even with this detail, some assumptions will usually have to be made:

- The size of your piles of snow
- The nature of your heart surgeries

Agile, Waterfall, and Project Documentation

Often, the conversation about the benefits of Agile and Waterfall methods resurfaces when it comes to documenting business requirements and project scope.

You may remember, back in Chapter 2, we discussed Agile and Waterfall. Agile is the project method where few requirements are decided upon up front, and the project team begins building software in close collaboration with the business. Frequent deliverables are provided and feedback is taken; so, in effect, the requirements are collected bit by bit.

The Waterfall method insists that requirements be stated up front in as much detail as possible.

Jumping right into a project and allowing requirements to be collected along the way is one of the most problematic aspects of Agile development. It can work in smaller projects and also in "from scratch" projects. Projects where the software has no dependencies on other systems are candidates for a from-the-start Agile approach with no or few up-front requirements.

But, in general, it is wise to get your high-level ideas in place and go through the murk to arrive at a set of final requirements. The requirements don't have to be as detailed as formal Waterfall might demand. You may simply point out your piles of snow and heart surgeries rather than defining them in detail. In fact, this is what modern project management ("agile enough") advises.

One critical rule to be aware of: In medium—to large-size projects involving one or more integrated systems and data migration, business requirements *must* exist in a moderate level of detail up front. To do otherwise risks the stability of the interconnected systems. The larger and more interconnected the systems, the more "anti-Agile" projects will be. Cyber-security considerations also require significant up-front planning.

The Scope Document

With your initial plan for the plan and subsequent breakout sessions, you will have ample material to create your scope document. Simply put, a scope document is a statement of the project "deliverables," what the project team will deliver to the business when the project is complete.

Here are the components of a good project scope document:

- Project summary
- Project deliverables
- Out of scope
- Constraints
- Assumptions
- Risks
- Timeline
- Budget
- Success metrics

Project Summary

You can usually take this from your project charter document. It's a paragraph or two on what the project is and why we are doing it, the business case for the project.

Project Deliverables

Your plan for the plan and breakout sessions will have identified the main project tracks. These are the "major deliverables." The breakout sessions will have unearthed the sort of work necessary to achieve the major deliverables. These are described one by one in the project deliverables section of the document. Where something requires further discovery (piles of snow and heart surgeries), the unknown aspect is noted along with the nature of the further discovery needed.

Out of Scope

Your scope document says what you're doing, and it also says what you're not doing. Some of these items will have emerged from your breakout sessions. Some things, it will become clear, are technologically impossible or inadvisable. Other items will not be possible given budget/timeline realities.

More about the interplay between timeline, budget, and scope will be discussed later.

Constraints

The world is full of constraints. The number of hours in a day is a constraint we all must work within! But, for the purpose of a software project, a "constraint" can best be understood as a fixed object (aside from budget and timeline) that is not going to change.

For example, the business may do work with the state government. The state must receive invoices in a particular way. Since the state is not likely to change their way of doing business, this is a project constraint.

Assumptions

We noted earlier that we might not know everything about the project, even after our final discovery, breakout sessions, and brainstorming. Some piles of snow and heart surgeries were identified but not explored in detail because that would take too much time. In those cases, some assumptions were made, such as, "The pile of snow won't be any bigger than X," or, "The heart surgery won't be any more complex than Y."

If you have made assumptions like that, you list them in your scope document. Here are some other examples of assumptions:

- That a key customer will accept the project's delivery date
- That a key vendor can deliver their part by a certain date
- That major project decisions and assumptions are not overturned by other business decisions
- That a piece of technology will work as advertised

Risks

A full discussion of risks is coming up in Chapter 8. In the risks section of your document, you will list all the major ones. You may have already guessed that this is the place to list your heart surgeries.

Risk lists almost always include:

- Dependencies on third parties you can't control
- An invariable date by which the project must be delivered
- Heart surgeries

Timeline

There are two basic ways to create a timeline: from the bottom up or from the top down.

I used an example earlier where a group of salespeople had some requests from a B2B company's customers. These customers wanted some new features on the website. This is a simple example of a case where a bottom-up timeline can be created. The programming team simply estimates how long it will take to complete the list.

In my experience, top-down timelining is much more common in software development. Companies usually have a set date in mind by which they wish to have the software complete. Usually the date is tied to some specific goal. A perfect example of this might be the need to launch a new product for a major conference. An e-commerce site might need to be up and running by Christmas. An accounting system might need to be in place before the start of the new fiscal year.

All of these are examples of software projects with fixed deliverable dates. In all of these cases, it is necessary to "back up" the timeline from the date of delivery.

For example, say the major conference is in March. The software team knows the system will need at least a month of QA (quality assurance) testing. Testing can't happen until all the data is in the system. So that means data migration must happen by February. The major business stakeholders need to approve that the site meets their expectations. This is called "user acceptance testing" or UAT. It looks like that will need to happen in January with a sample set of data. All this means the software needs to be delivered, ready for UAT, by the end of December.

That's how top-down timelining goes. You start with a known date and back up. At this point you won't be surprised to know that top-down timelining affects scope and budget.

A top-down timeline usually looks something like Figure 6.1.

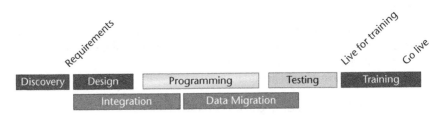

Figure 6.1

Budget, Scope, Timelining, and Horse-Trading

We'll talk in detail about budget in the next chapter. Just as with the timeline, as you create your scope document, it may become evident that certain things will fall out of scope because there is just no way to accomplish those things in the time and budget allowed. These things reveal themselves in the final discovery and business requirements gathering. It's normal and it leads to horse-trading.

Of course, it is not up to the project team to decide on what's in and out of scope. The scope document, therefore, is a kind of proposal to the business. It's the project team's best guess and judgment about what should and should not be in scope based on their understanding of the project goals, the timeline, and the budget.

When the scope document is initially presented to the business, the horse-trading will begin. Businesspeople will be upset that something has fallen out of scope. They may be willing to give up on some other thing to get the preferred feature.

Metrics

Your project charter document will have touched on this topic. The project charter establishes the business case for the project, often termed the "return on investment" (ROI). The business case varies project by project, but, in general, a company expects to earn some money or save some money by doing the project.

A high-level statement of business case or ROI, however, is different from a "metric." Metrics are also called "key performance indicators" (KPIs).

Let's take the integration project described earlier. Why would a company do such a project? The business case, in this instance, was that better management of membership data would allow the company to have insights and increase members. Now, how are we going to measure that? Of course, one metric is an increase in members. But how do we know that the increase came from the project? Further, there might be smaller things you would want to measure along the way such as whether more people filled out a form indicating an interest in membership.

Take another kind of project. Maybe your accounting system is out of date and soon to become unsupported by the manufacturer. That is surely a good reason for replacing it. But it has nothing to do with a success metric. In this instance, your business case would be different from the "more members" example in the previous paragraph. The business case would be stated in terms of the harms and loss the company might endure if they did not replace the accounting system.

Still, the question remains as to how the company will know if the project is successful. An answer in the accounting example might be to demonstrate at the end of the project that the new system performs all the functions of the old system. You can verify that with a checklist and interviewing users.

You might also want to know if the new system performs those functions better or faster and if it introduces new functions that save people time. Those would be great questions. And you would need to come up with methods ahead of time to answer them, such as a list of the old system's functions, but also current user perceptions about ease of use of the old system. You would interview users after the new system is finished to get their ease-of-use perceptions about the new system.

In general, success metrics say what the world will look like when the project is done, and how that differs from present reality in a concrete measureable way. What sales numbers are expected to increase? How many new customers will the project attract? What time or money is expected to be saved over what period? What adverse situation is avoided? The answers to questions such as these constitute your success metrics.

What About "the List"?

The breakout sessions are complete, the project scope document has been composed, and the horse-trading is finished. Next, the project manager, along with the programming team, creates all the individual tasks that must be done. Often those tasks are entered into something called an "issue tracking system." Basecamp, AtTask, and Jira all provide issue tracking. These granular-level tasks may then be assigned, completed, reviewed, debugged, and marked done. No wonder the PM has been eager to get to this part of the project. It's concrete, and it's where the software programming gets accomplished!

Defining and Visualizing and Project Scope

It is absolutely essential to be as visual as you can when you are scoping your software project. Draw on the whiteboard, draw on napkins—write on the table if you have to. Provide diagrams and sample screens to accompany your scope. I cannot emphasize this point strongly enough. Clearly defining and visualizing the project is arduous and it takes time, but it is one of the most important things you can do to ensure the success of your project.

Let me tell you what normally happens, so you can guard against it. Then I will tell you what it will look and feel like if you're following my advice.

What Usually Happens

You may have deduced from the previous chapter that at least half your software development team consists of concrete, black-and-white thinkers, and that, as a general rule, the business stakeholders are big-picture thinkers. This is a witch's brew that generally results in the following:

- Businesspeople attend initial meetings where the software development project is discussed. The meetings seem productive—fun, even!—as the possibilities of the software are outlined.
- The business analyst collects all the requirements in breakout sessions. The programmers weigh in and some back and forth occurs.
- A scope document is produced: words, words, and more words, specifying how the software will work.
- The programmers are pleased. They are seeing a picture take shape.
- The project manager is pleased. She will soon have her list.
- The businesspeople read the software requirements document. It may be the first time in their lives they have read such a document. It's really thorough with a lot of words, so it must be good, right? They sign off on what it says without fully understanding it.
- The project unfolds, and at the 60 percent mark, everyone realizes the business stakeholders aren't happy. Either they didn't understand the implication of something in the requirements document, or they understood but never thought it would be manifested in the way it turned out. Their complaints often begin with the phrase, "But we said in the meeting...."
- Project manager, programmers, and other software development people remind the business stakeholders that they *signed off on the document.*
- Chaos and conflict ensures. Often the vendor is fired.

Remember when I said in Chapter 1 that everyone would have a different idea of what the software development project would look like? That's what's happening now. It is absolutely essential, before this happens, to get everything down on paper—but I don't mean in words. I mean in pictures.

I have heard it said, "Well, so-and-so is a visual learner." The truth is *everyone* is a visual learner when it comes to software.

The Chicken and the Egg

What happens at this point is a classic chicken-and-egg problem. Your project team wants the business to sign off on final scope before putting pen to paper on screens and other visuals. But the business stakeholders need to see what the software will look like so they can confirm what they want.

You must get out of this conundrum—and fast. The only way is to hold your nose and jump. Get someone—anyone—to start mocking up screens. You will face resistance here on three fronts.

Barrier number 1—budget: The first area of resistance is budgetary. Visual creation is extraordinarily time consuming, so many people try to keep their powder dry because they know they will need to use it later.

HINT: PLAN FOR DEMOS

Budget in lots of time for mockups and demos. They will invariably pay off later.

Barrier number 2—resistance from the visual people: You will be tapping a resource who has visual skills. That is usually your UX or design person, but it is sometimes a good BA. Anyone with solid PhotoShop, Visio, or even PowerPoint skills will do. I feel so strongly about this that I myself spend half my time in PhotoShop and PowerPoint, despite the fact that I am an executive and team lead.

The visual person will wail, "But I don't know what to draw! I don't even have the wireframes." The only thing you can do at this point is to make assumptions and hope you're not too far off. Spend the time and do a mockup.

Barrier number 3—the programmers: One of the most effective strategies to achieve clarity is to have your visual people collaborate with your front-end programmer to make a realistic demo. Sometimes I even get the back-end programmer involved, coming up with mock datasets to populate the pretend screens. Your programmers will wail, "Wait! You want me to basically *do* the software before I *do* the software?"

Your answer is, "Yes. That's what I want you to do."

The truth is, we can all sit around and wish that business stakeholders could read a business requirements document and understand it, and that we would all have the same interpretation of what it's going to mean in the end. But as someone who has been on the front line of software development for 20 years, I am here to tell you that this never happens—literally, not once in my entire career. Projects limp along, and only later when the software is 70 percent done do stakeholders actually see and understand the essence of the project. They react and give feedback at this stage—which is often too late—driving costs up and blowing timelines.

Common Questions

Question #1: "How does this happen? I thought we hired a user experience person who will do the wireframes?"

That is true. But, as outlined earlier, generally wireframes come after a project scope is signed off. Then wireframes trickle in bit by bit. They don't form a coherent picture for the business stakeholders until too late.

Further, you may find out that the folks doing the formal wireframes didn't accurately understand the business requirements either.

Question #2: "Is this Scrum or Agile methodology you're talking about?"

Yes and no. As discussed in the Introduction, this approach employs some aspects of the Agile methodology—the aspects that, in my opinion, are most important to the success of a software project—while leaving out other aspects that may not be possible in most corporate settings. The key is to aim for storyboards, PowerPoints, screen mockups, and demos that capture large swaths of the project, so the business stakeholders can get a good grasp of the thing as a whole.

The key goals here are to create mockups that are:

1. Highly visual
2. Broad in scope

For large projects, these demos and mockups can happen in chunks.

I will share with you one of my most cherished compliments—the time a programmer said, "Anna, we really hate when you make us do this. It seems like such a waste of time … but it always pays off in the end."

Where Does Design Fit In?

You may be assuming from what I've said that I am asking for refined visuals. Nothing could be farther from the truth. In fact, I prefer to sideline the design team until very late in the project. At this stage, I prefer black-and-white visuals. I purposely avoid color because I don't want folks to start discussing color choice rather than functionality. Design is an utterly meaningless distraction until your business requirements are settled. In the demo process, business requirements will be naturally refined and vetted out as people see how the product will actually work.

It may all feel very fluid and disorganized to do it this way, but it is the most effective way to get everyone on the same page and head off late-stage problems.

> ## HINT: BETTER SOONER THAN LATER
>
> Late-stage problems are, by definition, always more expensive to solve than early-stage problems.

Working with Marketing Stakeholders

Marketing people tend to be some of the most highly visual individuals in an organization. They can sometimes have a very low tolerance for looking at something in a boxy, black-and-white stage. They really want to see how it is "actually going to look." You know you have a barrier here if you have a marketing person saying in a deflated tone, "Is that what it's *really* going to look like?" or "Where is the product shot going to go?" And as you may have discovered, often a CEO has come up through the sales and marketing channels, so he or she may be asking identical questions.

You need to reduce the marketing person's anxiety at this stage, or she will not be a productive participant. For this reason, it is often worthwhile to take a screen or two and send it to a designer for interpretation. That will show the marketing or other highly visual person that there are lots of exciting possibilities for design—and no, it's not going to actually look like that.

Once you have sufficient demos and mockups, you will have a much more solid feeling that everyone is on the same page. Business requirements can quickly get finalized, and you can move on to starting the project in earnest.

How You Know You're On the Wrong Track

You will always know you're on the wrong track at the planning stage of the project by looking for three telltale signs:

Too much talk: This means too many meetings about what the software will do. Having meetings—followed by e-mails repeating what people said in the meetings—won't help achieve a common understanding and vision.

Not enough visuals: As I have said before, you need screenshots, mockups, and demos. There is simply no substitute for these kinds of artifacts.

Too much continued excitement: Counterintuitively, if you're doing things correctly at this stage—with lots of visuals and demos—the excitement level of the stakeholders will and should drop significantly.

Years ago I read a piece written by a novelist who described what it was like to go from an idea to an actual work. She said when her concept was in the idea stage it was as if she was surrounded by a swarm of brightly colored butterflies. Then, when it came to actually doing the thing—getting words on the page—it felt like pinning the butterflies to cork.

When ideas are just ideas—tossed about in meeting rooms, swirling around in your own head—they are incredibly animated and exciting. Their real-life manifestation is, inevitably, less so.

It is human nature to want to stay in the butterfly stage. I think that's why so many of us are procrastinators. We somehow know that the result of our work will be less exciting once it is actually finished. So as visuals and demos are getting done, if you are doing this right, it will and should feel as though the air is coming out of the balloon. People may even say, "Is that *all?* I thought there would be more."

This is healthy and good. Your project needs to come down to earth.

A Word About Ongoing Discovery

Ongoing discovery!? We just had "final discovery"! What's this about discovery continuing?

One of the most anxiety-provoking observations I make to clients is that discovery will continue to the end of the project. This is a very bitter pill to swallow, but one that we must somehow choke down if we are ever going to get a handle on the reality of software development.

Many businesses and software people (BAs in particular) try to avoid this hard truth by doing herculean amounts of discovery up front, in colossal detail. It simply doesn't work. For one thing, generations of software developers have tried it in the traditional Waterfall method, and the results just aren't great.

There's another truth we must face: Human beings can only absorb and digest so much information. Understanding occurs in layers over time. People understand, and then they must "re-understand" and then they must "re-re-understand." There is just no practical way to beat out all the business requirements up front. Yet some seem to think if they only thrash hard enough, the wheat will separate from the chaff. Even if it were possible, most clients do not have the patience for it.

I've outlined my preferred strategy: Lots of visuals and demos to accompany documentation to arrive at good general agreement on features and functionality.

We simply must accept as givens the phenomena of continued discovery and of "re-re-re-understanding" and move on.

CHAPTER 7

Budgeting: The Budgeting Methods; Comparative, Bottom-Up, Top-Down, and Blends; Accurate Estimating

B udgeting is the dark, skeleton-filled closet of software development. When a group of project managers and account managers talk privately about budgeting, the conversation sounds a lot like a group therapy session, with one person after another admitting a sense of complete helplessness in the face of accurate budgeting.

Remember those software development theory books I referenced in the Introduction? I said there were lots of books available on theories of project management, and so forth. If you look to those books to learn how to actually budget a project, you will be sorely disappointed: The chapters are either very thin or missing entirely. Or sometimes it's the reverse: Some instructions on budgeting practically require a PhD to understand—evaluating programming tasks in detail, assigning "coding complexity factors" and so forth, and formulas for determining how much, exactly, your project is off budget.

Budgeting is another activity that *seems* hard because it *is* hard. That said, I do have some practical advice that will help you through the process and give you a chance to reach an accurate (if sobering) budget. Furthermore, I will disclose my team's own private budgeting formulas, which have worked well for us.

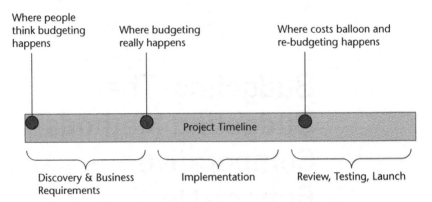

Figure 7.1

An Unpleasant Picture

Figure 7.1 captures the way budgeting happens in a software development project. Many people don't like what this picture says, but it reflects reality. Once we accept reality, we can create useful strategies for how to deal with it.

The timeline illustrates a common misunderstanding about software development among businesspeople and, frankly, a common delusion among software development teams who continually wish things were different. Budgets cannot be accurate until a solid amount of discovery and analysis has been performed.

What Goes on Behind the Scenes; a Scene

A client goes to a software development shop, in this case a web vendor, and outlines his desire for a website. He is attracted to the vendor because he admires their work. He has a brief list of requirements—probably no more than three to five pages.

Here is what happens at the vendor:

Account manager: Mr. Klientokos wants a website. Here's his request. (Hands over three to five pages.)

Business analyst: Hmmm, seems like this website is almost the same as Mr. Oteros' website.

Project manager: Yeah. I figure it would be about the same size.

Account manager: That's good. We were really off-base when we did Mr. Oteros' site. Remember that? We had to take a hit and so did he. Boy, was he mad at us!

Business analyst: Yeah. I remember. That wasn't pretty.

Account manager: So what did that site *really* cost? Can we pull the numbers on that?

Business analyst: Sure (clickety-clack on the keyboard). Yep. It was about $65,000.

Account manager: Great. I'll tell him $65,000.

Project manager: (reading the three to five pages): Waaait a minute. I see something here that's real science fiction. He wants some kind of recommendation engine to suggest products to his customers. All these other widgets he's requesting were on the Oteros website. But I can't even understand what he means by this recommendation thing. Better add on 30 percent for sci-fi.

Account manager: You got it.

END SCENE

This kind of budgeting occurs all the time in software development. It's a classic example of the first kind of budgeting, called "comparative budgeting."

Budgeting Type 1: Comparative Budgeting

As the name would suggest, comparative budgeting is when the software developers can use a project, recently done, as a comparison point. The software development team can be internal or external. The key element is they have hands-on experience with something similar to what is being requested. In the real world, this happens a lot in software development shops where the team does the same kind of project over and over for different customers. An example of this might be the implementation of an accounting system or e-commerce website development.

There is absolutely nothing wrong with this strategy. It works, and everyone gets a fair shake.

The key elements to successful up-front budgeting are:

- A relatively small project ($150,000 is probably the limit if the projects are *very* similar)
- A well understood kind of product, such as a website with common features
- The ability of the software team to accurately equate the requested project with another recently completed project

If you are thinking this sounds like the editing vs. creating topic, you're right! We're looking at a software project that is essentially an editing exercise: using what was done for Client A to know what needs to be done for Client B. Doing so limits the number of variables and, as we've discussed, risk, resulting in an easier-to-visualize and thus easier-to-budget project.

Gotchas with Comparative Budgeting

The pitfalls in the aforementioned budgeting strategy are pretty evident. In the previous scene, the vendor is making a *guess* that Mr. Klientokos wants what Mr. Oteros has. As described, Mr. Klientokos was attracted to the vendor because he already admired their work. So, in this case, the guess will probably be fine.

But what if they were wrong? The discussion between the account manager, BA, and PM occurred in private. No one went to Mr. Klientokos and said, "Hey listen, Mr. K., budgeting in software is really hard. No one can do it accurately. So here's the deal: We think your website is basically the same as Mr. Oteros' site. We're basing all our assumptions on that. If we're wrong in any way, the budget number won't be right."

Of course, no one really ever says this.

Here are some common gotchas in the comparative budgeting strategy:

- The equivalency that was made (project A is going to be basically the same as project B was) turns out to be inaccurate.
- A different technology is required. The project is indeed very similar, but the vendor works in open source and the client's requirement is for .NET.
- Areas of real risk (in the example, the "recommendation engine") are not spotted or properly understood.

Visual design issues can lead you far astray in making your equivalency decision. You may be attracted to something that *looks like* what you want in the visual sense, but it may not *do* what you want in the functionality sense. Functionality is where costs balloon.

Another mistake people make is in thinking that they must use a vendor that has worked in their industry before to get the comparison. In a small minority of cases this is a good strategy—if, for example, your industry requires certain functionality that is quite specific. However, I have found that many people only *think* their industry is special, with specific needs that can only be addressed by someone who understands their niche. This often turns out to be an exaggerated sense of uniqueness. It is much better to think in terms of *similarity* of functionality than to go searching for "the company that does websites for law firms."

Budgeting Type 2: Bottom-Up Budgeting

You may remember in Chapter 6, I introduced an example of a sales team in a business-to-business company. That team had requests from current customers and some potential customers about new features on the company's website. The list was handed over to the project team, which estimated the time it would take to complete each item (more on accurate estimating later).

This is a classic example of bottom-up budgeting. Bottom-up budgeting simply means going item by item on the list and giving a time and a cost to complete each one.

The Rub in Bottom-Up Budgeting

Having read the previous chapter and been introduced to the murk, you can probably already guess the problem in bottom-up budgeting: You need the list. More rigidly traditional Waterfall budgeters will insist on business requirements that are detailed enough to produce a definitive list and rely exclusively on bottom-up budgeting.

But, you'll also remember that our "agile enough" business requirements document did not go all the way down to the detailed list of each atomic task the software team would need to tackle. In fact, we even left in some assumptions around "piles of snow" and "heart surgeries."

Therefore, it may not be possible for you to do bottom-up budgeting on your software project. Or there may be large pieces of it where bottom-up is not possible. Obviously, if there is a detailed list around some aspect of the project, it can be bottom-up budgeted.

The key elements to successful bottom-up budgeting are:

- A clear list of business requirements
- A list that is down to the "task" level or close—feature by feature

Budgeting Type 3: Top-Down and Blends

In most software projects, there will be a significant number of unique requirements, lack of full lists, and other unknowns.

Here's how you know your project falls into this third budgeting scenario.

- *Size*: It is rare that a larger software project, in all its parts, can be equated to another project. The larger the project, the less likely you will have a fully detailed list, and the more unknowns you will have.
- *Integrations*: Most companies store data in specific, sometimes quirky, ways. A software project that must deal with a variety of legacy systems that are unique to the company will have budget challenges. It will be hard to find comparisons or to know the list.

■ *Features*: Any project that has unique or innovative features and functionality is by definition going to be tougher to budget.

This final point—functional uniqueness—deserves a moment. I have talked a lot in this book about the perils of creation, unknowns, and heart surgeries.

However, as I have also mentioned, often the *very reason* people want to develop a piece of software is to create something new that has not been done before.

What we're talking about is *innovation*. A business may want to deliver a product on the web in an innovative way, or automate a process to produce a competitive advantage. It is one of the most basic functions of not only software but technology in general—it drives innovation. Many times it's why people initiate software development projects in the first place.

Therefore, it may be highly inappropriate in many instances to strive to avoid creation. Creation may be the whole enchilada.

Top-down budgeting and blended budgeting are used for software projects with elements that are large, integrated, and innovative. Top-down, perhaps obviously, is the inverse of bottom-up. In practice, it means budget estimates are done holistically, not item by item. This is usually because you don't actually *have* the list of items to work from.

In top-down budgeting large chunks of the project scope—the big buckets—are estimated, usually in terms of weeks and sometimes months. As you might imagine, where equivalencies can be made, they are. This is where the blending comes in. Type 1, comparison budgeting, is brought into play. Further, where lists of things exist, they are estimated one by one. The team discusses best-case and worst-case scenarios around the piles of snow and heart surgeries. Frequently a range, high and low, is provided to correspond with best and worst.

It can take an experienced software team to do this kind of estimating. They need to be flexible and have practice in discerning the size of buckets. A software engineer who is accustomed to receiving the list may not be able to do this kind of budget estimating.

Why RFPs Don't Work

Many businesses send around requests for proposal (RFPs), soliciting bids on software projects. They figure if they can get enough bids, they will better understand the size of the project and what the budget is likely to be.

Most are stunned to see such a range of prices come in. Literally, based on the same RFP, proposals can range from $25,000 to $250,000. A client may think, "Good grief! I solicited RFPs to build an entire factory, and they weren't *this* widely different!"

Clearly, such a range is useless.

RFPs are not a great business practice, in my opinion, in many situations, but they are particularly nefarious in software. Even the most detailed RFP (say, 40 pages) will not capture everything. Each software vendor will tackle it differently. Some will try to do bottom-up, some top-down, some blending. So, even with a very thorough RFP, a software company is estimating largely in the dark.

Here's another scene:

Account manager: (reading the RFP): So, I have this RFP for a full publishing system replacement and website rollout. How long is that going to take?

Business analyst: I'm guessing discovery will be about eight weeks. They seem to have a lot of documentation. That's good.

Account manager: Okay, eight weeks for discovery. How long for the whole project? What's it going to cost?

Business analyst: Wow. Who knows what we're going to find in discovery? Isn't there a black-box legacy system we need to deal with? *Hmm*, the project feels like six to nine months. That's what most software projects are, right?

Account manager: We better say $450,000. That should cover us for things we discover.

Business analyst: Let's hope.

You can see all the vagaries and "guesstimation". Bringing the programmers in at this stage to read the RFP and come up with estimates gets no better results. First, as mentioned earlier, clients may *think* RFPs are specific—"It's 40 pages for crying out loud!"—but 40 pages is usually not enough, or it's not the right 40 pages for a programmer to be able to solidly estimate what the project will require. Programmers get very frustrated making commitments about gray areas where there are too many assumptions. They'll make a best-guess estimate, but they'll largely do it exactly the same way the BA in the previous scene did: "*Hmm*, feels like nine months to me."

Accurate Estimating and Comparison Budgeting

Getting accurate estimates using comparison budgeting is the easiest scenario. You need to get a previous budget for an equivalent project out of the drawer. Make sure the budget was accurate!

Scour the new project requirements for things that are different from the old project. These items may be list-friendly. Or they may be buckets. If there are differences, sometimes called "deltas" in software parlance, you'll have to account for those.

Otherwise, you are simply looking at comparative size. If the project is double the size, you begin by doubling the budget. Yes, you do get efficiencies as a project grows, but larger projects also become unwieldy. You'll want to make sure you give yourself enough room.

Effective Estimating in Top-Down and Bottom-Up Budgeting

Here is the method my team and I use to budget as accurately as possible. It's a budgeting method to match the "agile enough" approach. There is a little math coming, so hang on to your seat.

Establish a Base Budget for Programming, Ongoing Discovery, Unit Testing, Debugging, and Project Management

All software budgeting must start with a base, and the base is programming and associated tasks. Your base budget for software development includes:

- *Programming time*: This is the number of hours (or weeks) a programmer estimates a certain task (or bucket) will take.
- *Ongoing discovery*: This is the amount of additional discussion the programmer will need to fully clarify the item.
- *Unit testing and debugging*: When a programmer is done with an item or feature, she must spend some time making sure it works. This is called unit testing. Further, when formal QA happens (different from unit testing), the programmer will have to do additional debugging on the item. That's all included here.
- *Project management*: It takes a certain amount of effort for the project manager to track all the tasks, assign them out, have daily meetings, and confer with programmers.

Percentages of Each

A well-run software project usually has the following percentages of each item:

- Programming features: 60 percent
- Additional discovery: 15 percent
- Unit testing and debugging: 10 percent
- Project management: 15 percent

Therefore, if we were talking about a programming effort of 100 hours, it would look like this:

- Programming: 60 hours
- Additional discovery: 15 hours
- Unit testing and debugging: 10 hours
- Project management: 15 hours

Programming Hours—Raw and Final

When a programmer tells me it will take 10 hours to do something, I call those the "raw" hours. I then increase that number by 40 percent to get the final hour estimate. I have learned to do this in my career because programmers are, as a group, optimistic people! They usually underestimate the amount of time something will take.

Furthermore, the programmer trait of black-and-white thinking comes into play. It may indeed take 10 hours to code the item. But there are bathroom breaks needed, and pencil chewing, and the set up of development environments and other things. The programmer may be accurate it will take 10 hours of typing on the keyboard to literally code the feature, but he's missed everything else. A general 40 percent seems to cover it.

Of course, with a well-known software team, different percentages may apply to different team members. I do know some programmers who are more accurate than others.

Bottom line: When a programmer tells you 10 hours, make it 14.

The Math Part

Why do we have to do math? That 60-15-10-15 percent cited earlier looks so clear! It all adds up easily!

The reason you must do some math is that you *start with the programmers' estimates*. You don't start with the whole picture. Let's step through a real-world scenario so I can show you how it works. We'll do it with one item.

We have a feature; let's say it's the implementation of a new website, and a programmer tells you it will take 50 hours.

We immediately increase that number by 40 percent for bathroom breaks, pencil chewing, development environment setup, and human error: so 70 hours.

We know 70 hours represents 60 percent of the total project. How do we derive the other numbers? (Trust me and your high school math teacher. You can't multiply the 70 hours by 10 percent and 15 percent and 15 percent and see what comes out.)

The math problem is as follows:

Seventy hours represents 60 percent of *what number*?

Put algebraically, we have

$$70 = .6 * X$$

Solving for X we get

$$70 \div .6 = 116.66$$

Rounding up, we call that sum 117 hours.
We now know that *117 hours represents the whole*. Now we can do the math for the other elements:

- Additional discovery: 117 * 15 percent= 17.5 hours
- Unit Testing & Debugging: 117 * 10 percent = 11.7 hours
- Project Management: 117 * 15 percent = 17.5 hours

You might ask, "Why didn't you just call it quits after you knew the whole project was 117 hours? Take 117, multiply by a billing rate, say $100/hour, and your project is $11,700."

You could indeed do that if all the resources on the project have the same billing rate. But typically they don't. Additional discovery is usually done by a combination of business analyst and programmer. Therefore a blended rate may be necessary. Project management is a different resource entirely.

You can do this kind of budgeting for different features or buckets, even if multiple people work on a feature. Further, you can adjust percentages—discovery and project management in particular—to suit your project. Maybe you feel your project will be light on discovery but higher on project management. Then you would shift the percentages in favor of discovery. Sometimes a project is virtually all programming with very little business stakeholder input or need for communication. If that's the case, then you might state programming hours represent 70 to 80 percent of the total project.

My team sometimes debates about the 15 percent discovery factor applied to all programming items. In truth, some of those items will be well understood. However, other items will be particularly poorly understood. For example, we may know we have to on-board the data from that black-box legacy system, but no one has worked with that system since the 1960s. So

Project Budget	Programmer 1 Backend	Programmer 2 Front End	Total Raw Hours (+40%)	Total Size if Programming is 60%	Discovery	Unit Testing	Project Management	Programming ($150/hr)	Discovery ($135/hr)	Unit Testing ($150/hr)	Project Management ($125/hr)	Totals
Feature 1	7	0	9.8	16	2.45	1.63	2.45	$ 1,470.00	$ 330.75	$ 245.00	$ 306.25	$ 2,352.00
Feature 2	2	4	8.4	14	2.1	1.40	2.1	$ 1,260.00	$ 283.50	$ 210.00	$ 262.50	$ 2,016.00
Feature 3	0	8	11.2	19	2.8	1.87	2.8	$ 1,680.00	$ 378.00	$ 280.00	$ 350.00	$ 2,688.00
TOTAL												$ 7,056.00

Figure 7.2

who knows what we're facing? Further, as features are released, the business stakeholders may come back and say, "Wait a minute! That wasn't what I envisioned." Or possibly, "*Um*, I hate to tell you this, but we changed our minds." A blanket 15 percent seems to cover it in our experience.

Figure 7.2 shows what this kind of budget looks like. You can find this spreadsheet with formulas already there at my website (http://emediaweb.com/completesoftwarepm).

That's it. This is the method we use to get to a good base budget before adding other known costs in and considering risk.

We'll talk about risk in Chapter 8.

Additional Items to Consider

Almost all software projects have costs in addition to the actual programming and associated tasks such as project management. The trick with these items is to make sure you've thought of them all. For example, it often happens someone says, "Shoot! We forgot that we'll need a server upgrade to accommodate the new software! We didn't estimate that!" Or, "Nancy is going to need Photoshop in order to edit pictures and put them in the new Digital Asset Management System (DAM). Did anyone think of that? How much is a Photoshop license?"

Once these items are on the table, it's pretty easy to get costs. Here's a list you can work from:

- *Hosting costs*: The fees a hosting company will charge to keep a website running, for example.
- *One-time setup fees*: Other fees charged by the hosting company, for example.
- *Hardware*: You may need to buy hardware on which your software will run, a server, for example. Other times people in your organization may need to upgrade their hardware to run the software.
- *Software*: People in your business may need to install or upgrade other software as a consequence of the project: for example, if they need a newer version of MS Word to work with the new software.
- *Licensing fees*: This applies to software, but also to system-level technology such as server software, licenses to project management

software, metrics software, ad-serving software, or any third-party software related to your project.

- *Subject-matter experts*: These are vertical experts who might be needed.
- *Quality assurance testing*: For medium- to large-size projects, it is usually advisable to hire an external, independent QA team (more about that in Chapter 11).
- *Training*: Your businesspeople may need to be trained on the new system.
- *Documentation*: Often needed to accompany training and to hand off the new system.
- *Customer service*: Your customer service people may need to be trained on the new system.
- *Communications, PR, advertising*: Your company may have to invest in these items if it is launching a new piece of technology to a large customer base.

All of these items must be captured, and the costs associated with them must be obtained. However, as noted, these items tend to be more concrete. They are not the costs that so famously go out of control on software projects. The more typical scenario with these items is "Oops. We forgot." In general, the programming effort is where cost ballooning takes place.

Budgeting and Conflicts

It'll be no surprise when I say conflicts often arise over money. Here are some common conflicts I see in the budgeting process:

- A software vendor has a very "waterfall" approach. They will not agree to a budget until they have a detailed list of requirements. Furthermore, they want an initial fee to do all the work to get to the requirements. The client says, "Can't you just give us a ballpark number?" The vendor says they can't.
- The budget is created in a top-down way, and midway into the project one of the "black boxes" is causing much more work than initially anticipated. The team goes to the business leadership and says they need more time and money. The business wonders how they got it so wrong.
- A software project is over time and over budget. The business leadership does not understand why this is happening. The last project came in just fine! They don't know the last project was done using the comparison budget method with a solid equivalency.

You can get insight into all of these conflicts and potentially avoid them if you are aware of what kind of budgeting is going on in your project. If you are working with a Waterfall-style vendor, asking them for a ballpark or "swag" budget is going to make them very uncomfortable. And, yes, it is customary to pay for tech discovery in order to achieve a very detailed list in this situation.

The benefit is that you usually arrive at a very solid number because there is so much detail. All of that is great. But there is a downside as we discussed in Chapter 2. The Waterfall-style process severely discourages shifting from the list. Make sure you are aware of these things and that your business's style and needs fit well with this Waterfall style of budgeting and execution.

If the budget is created in a top-down or blended method, on the other hand, it will be less solid by definition. The budget presented earlier bakes in time for additional discovery and shifts in features, which leads to greater accuracy. Still if you have some real heart surgeries and lots of innovation and black boxes, your budget is less solid. It's key to know this up front, and we'll talk about it in Chapter 8. On the business side, be a wise consumer of software services both from internal and external teams. Understand how the budgets are built. Since most medium- to large-size software projects use a blend of bottom-up, top-down, and comparative, it's useful to know the proportions of these methods in your budget.

I have heard literally dozens of arguments based on a basic misunderstanding of budgeting methods. An irate businessperson shakes her fist (metaphorically speaking) at a project team saying, "Why won't you agree to a number and stick to it!" She declares that team Q or vendor D did that in the past with no problem.

If team Q or vendor D were working with a good comparative project or really solid list, of course the businessperson had a stick-to-the-budget experience! Maybe this time it's different.

We're not quite finished with our budget yet, because we don't have a contingency factor. But first, we have to learn about risk.

CHAPTER 8

Project Risks: The Five Most Common Project Hazards and What to Do About Them; Budgeting and Risk

"Risk" is a very basic word in our lives. Driving a car is risky; riding a horse is risky; starting a new relationship is risky. Most people translate the word risk into the following statement: "There is a chance that I will endure some harm, but in all likelihood, I will not."

The word risk in software development has a totally different meaning. Software risks can be defined as the number of unknown and inherently "complex" elements of a project that have a *high likelihood* of impacting budget and timeline.

You may recall in Chapter 1 we discussed the difference between the simple, the complicated, and the complex. To review, simple tasks (e.g., shoveling piles of snow) are easy to understand and simply take effort—sometimes a lot of effort. Complicated tasks (e.g., assembling an Ikea desk) may take a lot of time, have lots of steps, involve mistakes and do-overs, but with enough diligence there is an almost 100 percent certainty that they will get done. In contrast, complex tasks (e.g., heart surgery on an elderly patient with several unknown underlying conditions) are inherently risky. No matter how skilled the surgeon is, there is no way of knowing if the patient will get off the table.

In terms of risk, ideal software projects would consist of lots of Ikea desk and snow shoveling activities and would involve very few heart surgeries. In life, though, sometimes you can't choose when or if a heart surgery needs to occur. However, if there are aspects of your project that will involve heart surgery–like operations, you both can and must know about it ahead of time.

So, you should evaluate your project based on

- *Pile of snow (very low risk)*: Arduous tasks that simply require labor; the likelihood of success is high as long as the labor is undertaken.
- *Ikea desk (low risk)*: Complicated tasks that require skill, diligence, and do-overs; the likelihood of success is also high, even if the effort is time-consuming.
- *Heart surgery (high risk)*: These are complex tasks where there is little ability to judge how long or how much effort the activity will take, or if it will be successful at all.

Sticking to your budget and timeline requires minimizing heart surgeries and maximizing desks and snow piles.

HINT: LISTEN FOR HEART SURGERIES

Many businesspeople get very anxious over the Ikea desks and piles of snow and not so much over heart surgeries. Sometimes it's just that the first two *sound* so intimidating (e.g., "We're going to have to enter data for two thousand products!"). A heart surgery might sound like, "We're just going to develop a custom piece of code that integrates the two systems."

Five Always-Risky Activities

In tackling software development projects, five items *always* go into my risk column until proven otherwise:

1. Integration
2. Data migration
3. Customization
4. Unproven technology/unproven team
5. Too-large projects

Integration

As we've discussed, when a software product has to exchange data with another system, this is called an "integration" or "an integration point." Here are some examples:

- An app that takes user feedback and publishes it to a separate website
- An e-commerce site that has to send orders into an accounting system
- A publishing tool that must push articles to both a print system and a web content-management system
- Anything where the words "web services," SOAP, or REST come up

It is a truism of software development that integrations can be difficult to execute and are hard to debug. They are, however, a reality of life. The ability of one system to exchange data with another provides the power and functionality most users have come to expect from the software they use. Therefore, it's not useful to view integrations as something to be avoided. Rather, it is important to be alert to the integration aspects of your software project and know that those aspects inevitably involve risk.

HINT: TOO MUCH INTEGRATION

Software developers can be "integration happy." Let's integrate this with that and that with the other and the other with the third. Integrations, a necessary reality of software development, should be minimized to a set of core needs and simplified to the extent possible.

Data Migration

Many software projects often involve remaking an existing system. Maybe you have an inventory tool that's 10 years old. Or an estimating system that someone designed back in the 1990s is still running in your office. Or maybe you want to re-launch your content-rich website. In all of these cases there is data in an old system you don't want to lose.

In order to get the new system up and running, the software development team is going to have to figure out how to export the legacy data out of the old system and import it into the new system. The older the data, the more difficult this will likely be. Challenges like mapping problems, dirty data, and missing data can all plague data migration activities. The new product that goes live appears buggy, and people cannot reach their data.

As with integration, migration is an inevitable part of many software projects. Rather than avoid it, you should anticipate and plan for problems that may arise.

Note: In previous chapters, I have referred to data migration as a "pile of snow" activity, not a heart surgery. And in many cases this is true. However, data migration "snow piles" are often so big and so unknown that I always put them into my risk column. Further, many data migrations require custom code to be written to automatically transfer the data from one system to another. In this case, it becomes more like a heart surgery.

Customization

Earlier in this book, we talked about the idea of editing versus creating. Calling something "custom development" is another way of phrasing the same idea.

Here are two common examples of situations where "custom" development arises:

- Company A has decided to use an e-commerce software package to launch its new site. The package has "out-of-the-box" page templates for the checkout process. But the product department of Company A wants to change the way the cart works. They need the ability to ship to multiple locations. The programming team hired to implement the software says it will be "no problem" to customize it for Company A.
- Company B has chosen a learning management system that does six functions out of the box. Unfortunately, Company B likes only four of the functions and really needs another one that is sort of like function six, but not really. The software developer tells Company B that it will be "no problem" to customize the software exactly as the company wishes.

Don't believe anyone who tells you it will be "no problem" to "customize" something. Customizations are by definition risky, heart surgery activities. The reason has to do with editing versus creating. If you are creating or inventing something new, you have no idea how long it will take you to get it done, or if you will succeed.

Sometimes a software developer will say something like, "We did a customization for Company C that is the same as the customization you are asking for. So we feel confident it can be done for X dollars and in Y timeframe." Signing up for a customization that has a track record is far less risky. In fact, because it has been done before, it's not really inventing anything, but rather adopting a previously created customization.

Note: In this scenario, your questions and investigation would focus on the implications for upgrade path as discussed in Chapter 3.

Unproven Technology/Unproven Team

Many times, techies within a company want to use a new or unproven technology. Say, for example, that a content management system has become "hot." Or maybe there is a new development language that is supposed to be the most powerful thing around.

I advise caution here. It is extremely risky to be a first adopter of a new or fairly new technology. Not enough projects have been executed using this new technology to even be able to identify what the problems with it might be. It's like having the first model year of a car.

Similarly, you might have a programming team that is inexperienced with a particular technology, especially if it's new. This is the same theme. They may be smart and have a great track record in PHP. That doesn't mean they can succeed in this complex Java project. Remember from Chapter 5. Software programmers are not C-3PO from *Star Wars*. Find someone who is an expert. Even with new and hot technologies, there is usually *somebody* who has a track record.

Obviously, if you have the two together, unproven technology with an unproven team, your risk profile soars.

Too-Large Project

The larger the project, the more inherently risky it will be. Projects that are very large can simply be too big to "get your head around." Often, a large project means separate teams will have to do pieces of it. Alternatively, one team may have to tackle it section by section. This kind of approach, necessary in a large project, is almost like an integration. One separately completed piece has to work seamlessly with another separately completed piece for the project to go live. One good alternative is to figure out how a very large project can be accomplished in phases.

Want Versus Need

To be clear, no one is saying that the risk areas such as integrations, data migration, and customization are "bad" or should be avoided at all costs. In many ways, these risks are a reality of modern software development life. However, they should be regarded with the appropriate level of seriousness. For example, all high-risk activities need to be carefully weighed in terms of want versus need.

Want Versus Need: Programmers

Unfortunately, many programmers and "techies" encourage business stake-holders to wade into the major risk areas of integration, customization, and unproven technology. In part, they do this because they want their lives to be interesting. Constantly sticking to out-of-the-box functionality is boring to a developer who looks for satisfaction in his job and accomplishments on his resume. Also, I have found developers to be quite overoptimistic about what they can do. It is all too common for developers to tell business stakeholders that such-and-such a customization will be "no problem."

Don't believe it.

HINT: BEWARE THE OVEROPTIMISTIC PROGRAMMER

Seasoned programmers who have been burned once or twice by over-promising tend to be more realistic than younger or more inexperienced ones. But *tend* is the operative word here. Many programmers carry the trait of over optimism with them throughout their careers.

Want Versus Need: Business Stakeholders

Many times business stakeholders become very demanding with regard to the risk areas outlined previously. They will tell you they *absolutely cannot live* with the out-of-the-box functionality as is. They must have such-and-such cus-tomization, integration, or data migrated into the system.

Unfortunately, they may have also been assured by the overoptimistic development staff that what they are asking for is "very doable" and "no prob-lem." Alternatively, they may "have a friend" who did "exactly the same thing" at Company D, and it was easy as pie. (Once again, do not believe it. Treat such assertions with the healthy dose of skepticism they deserve.)

Challenge business stakeholders to separate the "need" business require-ments from the "want" business requirements, especially when it comes to the five major risk areas.

Optimism Is Not Your Friend in Software Development

I use this phrase all the time: "Optimism is not your friend in software devel-opment." Be skeptical. Question. Ask for backup. Especially when it comes to the five major risk areas. In the hundreds of software projects I have led, it is rare that the optimistic "it will be no problem" scenario has panned out. Edit yourself and your staff when you sense optimism in the air rather than brass-tacks pragmatism.

Beware the Panacea Claim

We are all familiar with panacea claims. Things like, "Proven diet aid! Guaranteed to succeed where other diets fail!"

In the software world, panacea claims sound like, "It will run on any device." Or, "build once, run anywhere." Or, "integrates with any system." Or, "transparent data migration." Or, "works seamlessly out of the box."

At their core, claims like these promise to simplify the really hard parts of software development. If they were true, they would represent the holy grail of technology. Just like in every other aspect of life, if it sounds too good to be true....

Many panacea claims *do* have an element of truth in them. A product may indeed make something easier or smoother. "Works on any device" may actually mean "works on many devices." It may still be worth it to sign up for such a service, product, or technology. Just treat the panacea claim with skepticism.

Facing Risks

The following statements are true about risks in software projects:

- Many times, risks cannot be avoided.
- Risks cannot be fully neutralized, even if you are aware of them.
- If you add risky factors to a project, risk consequences (namely, extensions to time and budget) *will very likely* happen.
- Risk consequences always translate into more time and/or cost.
- Early-stage risks are by definition less costly than late-stage ones.
- Encountering the consequence of risk is no one's "fault."

A Few Words About Fault

As discussed in Chapter 1, if a surgeon does open-heart surgery on a person with an undiagnosed medical issue, problems might arise during the operation that the doctor cannot solve, and the patient could die. This is no one's fault. It is a consequence of engaging in a highly "complex" and "risky" activity.

The more risk you take on in a software development project, the more problems you will have. Not *might* have. *Will* have. It should be viewed as a one-to-one correspondence. Even if everyone on your team is the best, most prudent, most diligent professional in the world, more risk equals more problems. It is no one's "fault"—it's simply the nature of the beast. The only effective strategy for success is to understand how inherently risky your project is and then prepare for problems, making sure you have processes in place to deal with them as well as contingencies in the budget. (In later chapters, I will talk about dealing with problems and the specific strategies you will need to have in place to deal with them.)

Identifying Risks Up Front

"How inherently risky is my project?" It is critical that you not only ask but keep asking this question. Can you identify the risks? What about the five biggies?

You might ask, "Do the five biggies cover all of it?" The answer is: yes, mostly. But, of course, every business is different and every development project can have its own quirks.

This is where our pile of snow, Ikea desk, and heart surgery analogies come into play. You can identify the risks in any software project by using these concepts. As you discuss the features and functions of your project with your team, you will begin to be able to identify which ones sound more like heart surgery.

I am repeating these analogies again and again because of their importance to software development projects, even though they might sound a little silly. They work! All team members need to be schooled in telling business stakeholders what is a pile of snow, what is an Ikea desk, and what is heart surgery.

Embrace the Snow

As indicated in one of the earlier Hints, I have often experienced business stakeholders react with great fear and emotion around what is basically a pile of snow: "Do you realize that Greg is going to have to key in two thousand items by hand!"

Such an inflammatory statement can be enough to whip project members into immediate action. Programmers get busy writing custom code to rescue Greg from this (admittedly laborious) task. And then what has happened? You have traded an inherently un-risky pile of snow task (keying in 2,000 items) into an inherently risky one (custom coding a solution).

Ironically, business stakeholders seem to readily accept statements like "this customization is no problem" and panic over "two thousand data items to enter." To be fair to business stakeholders, overoptimistic programmers often present complex, heart surgery tasks in a benign-sounding way. They present snow piles in overwhelming and intimidating ways.

Do not fear piles of snow; they are your friends. Talk to the team: "How long will that take Greg? A week? Possibly two? Well, that's not so bad, is it? Do we really want to write custom code for something that will be done in a couple of weeks?"

As mentioned earlier, the "pile of snow" meriting the most concern is a large unknown mass of data requiring migration. A key reason for this is the desire or need to lessen "show shoveling" (by-hand data entry) through automation, such as custom coding a script that will take care of the migration process.

HINT: TRUTH AND CONSEQUENCES

Take as a given that all project risks *will* have consequences. Do not stick your head in the sand. Remember that a risk means a consequence, which means a problem that will cause a delay or cost money or both. Optimism that "it will all work out fine" is not your friend.

Talking to Your Boss

Unless you are a sole proprietor or top dog in an organization, you have to answer to someone, and it's best if that someone is brought into the discussion about risk.

In most areas of management, making something a requirement is usually enough to ensure it gets done. If a boss is serious enough about something—making threats or fist pounding where necessary—people will straighten up and fly right.

Often you will see many top bosses resort to these techniques at the *end* of software projects. The project is off schedule and over budget. A top manager may think, "Clearly people didn't understand that when I said July 1 and $75,000 I *meant* not a day later or a dollar more." The fist pounding ensues. But to no avail, because the "risk chickens are coming home to roost." It wasn't ever about diligence and effort.

Bring your boss—or whomever you are accountable to—into the discussion of risk at the very beginning. Share this chapter with him or her if necessary. Sometimes top bosses are actually able to remove risks that you are not able to. For example, a top boss may be able to insist on minimal "customization" and drive the organizational behavior change to make that a possibility. Or, a top manager can put the kibosh on some slick new technology that an influential cadre of people want—especially if he understands it will impact his timeline and budget in the end.

A final strategy is for the top bosses to assign a contingency number (time and budget) in proportion to the increased risk. Later in this chapter, I will get back to risk and assigning contingencies in budgets.

The most important thing is for senior business stakeholders to completely integrate the following into their thinking: Increased risk will mean increased time and budget. This is an invariable rule. Not might, but *will*.

Hidden Infections

We have spoken about five concrete project risks that are technological in nature and straightforward to identify. You can ask yourself, "Is my project large? Does it involve integrations and data migrations? Are there customizations? Am I using a new technology?"

There are three additional, sometimes less obvious, issues you must be aware of. Think of these as types of "diseases" that infect your project and that you want to diagnose and treat early if you can:

1. Bad technology team
2. Wrong technology choice
3. Too many opinions and lack of leadership

Bad Technology Team; Wrong Technology Choice

In Chapter 4, I remarked that team selection and tech choice happen best in concert. In my entire career I have never seen a software project succeed that had a bad team working with an inappropriate technology. As (bad) luck would have it, the two almost always seem to go together.

Too Many Opinions and Lack of Leadership

Further, if leadership lacks confidence and lets the "peanut gallery" get out of control (with everyone citing panacea claims, or the experience of their brother-in-law, or what was done successfully at Company D), projects dissolve into chaos and confusion.

What happens then? When it becomes clear that the project continues to consume time, money, and resources in a seemingly indefinite manner, executives have no choice but to take action. Sometimes a failed software project is high profile enough that the top managers who sponsored the project lose their jobs. A new leadership team comes in and starts from scratch. If you're lucky, you may just need to fire the technology team and find a new one. Usually the new one will need to throw everything out and start over.

In Chapter 12, we will look at how to address the problems caused by both risks and "infections."

With the many risks and potential pitfalls understood, we now have enough information under our belts to have a purposeful discussion about contingencies and finish up on budgeting.

The Contingency Factor

In Chapter 7, we went through the three methods of budgeting. We'll get back to budgeting now and how to work in contingencies.

It's pretty straightforward. First you must identify the risks you are accepting. Then, for the purposes of budgeting, you must assign a percentage to each risk. You add up all those percentages to get your contingency.

One basic rule: The contingency factor can never go above 200 percent, because that is the cost of throwing everything out and starting anew at a late stage of the project.

"Why 200 percent?" you may ask. "Why not 100 percent?"

The reason we say 200 percent is that many times the "risk chickens" only come home to roost in the last stage of the project. So the initial budget has been largely spent—if not overspent. If you have to throw everything out, you'll need to re-fund the entire project.

It's my intention to be sobering here. Very sobering. I find that both technical and nontechnical people do not adequately weigh the consequences of the risks they take onboard at the beginning of a project. Perhaps wishful thinking is an unavoidable human trait. Also, I've noted in previous chapters that the start of a software project has "honeymoon" qualities, with lots of excitement that can obscure practical realities. Whatever the case may be, it is an unfortunate fact that risks are frequently accepted without full understanding of their potential consequences.

The Cost of Consequences

I find it is helpful to talk about the "dollar value" of particular consequences.

Here's what usually happens in a discussion of possible consequences. A technical person says, "Well, if the integration with the state database doesn't unfold as planned, Peter will have to write a piece of middleware. Or perhaps we'll be lucky, and they will have updated their API as promised by then."

Table 8.1

Type of Risk	Suggested Contingency
Tricky data migration	+10%
Integration	+5% each
Large, "un-chunk-able" project	+15%
Unproven technology	+25–50%
Customization/innovation	Assigned in proportion to the amount of customization/innovation involved (e.g., a project with 20% innovative features means a +20% contingency)

To the business stakeholder, this sounds like, "*Blah blah blah* hydrocon-fabulator, *blah blah blah* qui-fi-triple-jammer."

To drive effective decision making, it's much more useful to say, "Look, we've decided to integrate with the state database, but there's no established protocol for that. We've put in an assumption of how it's going to work. But if that assumption proves false, it could cost $150,000 more. So, do you want to go ahead with this integration?" You'll definitely have everyone's attention at that point.

Contingency Percentage Factors

My firm uses a formula to get to the dollar value of consequences. It's a contingency percentage calculation you can perform when you make the decision to on-board risk.

Table 8.1 provides general rules of thumb for calculating contingencies.

Add up your risk factors, and that's your contingency. So, for example, a large, un-chunk-able project with two significant integration points and 5 percent new innovative features looks like this:

Budget based on one of the three methods:	$250,000
plus large project factor:	*$37,500*
plus integration factor:	*$25,000*
plus innovation factor:	*$12,500*
Total contingency:	*$75,000*
Total budget:	$325,000

In the Real World

Of course, in the real world this can look different. If you have some experience at this, you can assign percentage contingencies in project-specific ways. For example, one area that's more art than science has to do with the level of innovation that's going to occur in the project. For this, I need to look in the whites of the programming team's eyes and judge their level of confidence. Are they feeling confident that the innovation requested is well within their grasp? Or are they thinking, "This sounds like science fiction"?

That said, many of the percentages above are pretty reliable. For example, I can't imagine tackling a very large software project without setting aside a 15 percent contingency.

The Good News

Your contingency calculation is simply a way of assigning dollar value to risk. It's just a tool you can use in decision making. If you are concerned about the high levels of contingency needed, you might consider making some different decisions.

Take the earlier example. You may decide that one of your integrations is a want instead of a need. Mary can continue to export the spreadsheet and put it on the file server where it can be uploaded into the second system. Immediately deduct $12,500 from your contingency. Now the contingency has gone from $75,000 to $62,500 because you've decided to delay the integration.

A Common Question

"So, a contingency means a budget overrun that might happen. It also might not, right?"

In software development, risks usually have consequences. It is true that sometimes things work out. You dodge the proverbial bullet, and the "risk chickens" don't come home to roost. Unfortunately, this is rarer than we'd all like to hope. Also, what frequently happens is that you dodge the first bullet but there's another one coming that you didn't anticipate.

A common tech adage states that there are three factors governing all software development projects: (1) time; (2) budget; and (3) scope (e.g., features and their associated risks). You get to pick two. In other words, if you want your project to come in on time and on budget, features will have to be sacrificed. In my experience, organizations are *extremely* reluctant to sacrifice features.

Therefore, my advice is to always budget for your risk factors and be prepared for the contingency to be spent.

Long-Term Working Relationships and Contingency

In long-term working relationships (e.g., with a long-standing vendor or experienced internal team), it is not always necessary to assign contingencies to this degree, and you can usually stay closer to the base budget. The reason here is common sense. The longer people have worked together, the closer they will be—both consciously and unconsciously—in their overall assumptions and in communicating with each other effectively.

CHAPTER 9

Communication; Project Communication Strategy; from Project Kickoff to Daily Meetings

There's a common thread among all successful software projects: They have good communication up, down, and across (Figure 9.1). Information flows from management down to the team members, from the team members up to management, and then across among all team members.

Informal communication goes on constantly. Still, it is not sufficient to rely on informal communications, as many technology teams do. Formal communication structures are necessary. The bigger the project, the more are needed. Even though most people would like to avoid that one extra meeting, if the meetings are purposeful and regular, they will be invaluable for catching mistakes, miscommunications, and gotchas as well as for keeping the project moving.

Approaching your regular weekly meeting, you may feel challenged. What are we going to talk about? What should be on the agenda? Just the act of putting an agenda together will be a healthy exercise for the project. In my experience, regular meetings help nudge a project along and keep teams from spinning off into their own solar systems. Here are ways to structure your meetings as well as agendas that are good to follow.

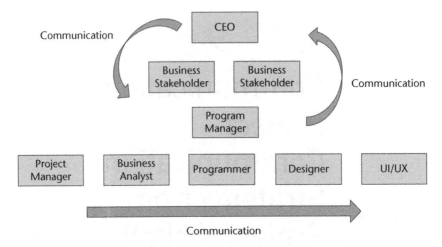

Figure 9.1

Project Kickoff

The kickoff is the communication piece most often left out. Everyone knows we're doing the project, right? Why don't we just get started? You may even feel a little silly calling a project kickoff meeting, like you're making a mountain out of a molehill.

Don't be one of those people who forgets the project kickoff! It's not just a rah-rah meeting to shake pom-poms as the project team does jazz hands. As you'll see next, the project kickoff has key elements that will help you later.

Project Kickoff Cast

The project kickoff must include the right people. Many times a project is so important the company president will call a company-wide meeting to announce it. If that happens, great. Still, I recommend a kickoff, separate from a company-wide announcement, with a tighter group. This group includes (1) the project leadership, and (2) company leadership.

Project Leadership

In Chapter 4, we discussed project leadership roles. They are summarized next in review.

Project Sponsor The project sponsor is the key executive in the business who has responsibility for the project. This is often a businessperson, though sometimes it may be a CIO. She is the main voice saying, "Hey, we need this project." Often, she is the one who must report to the CEO or board on the status of the project.

Program Manager/Account Manager This is a trusted technology professional, often a consultant, with deep experience in leading teams and projects.

Project Manager This person has charge of the project breakdown: all the day-to-day tracking tools including ticketing systems, wikis, and budget reporting tools. Formally trained in the methods of project management, this person will lead daily meetings with the programming team and other teams.

Team Leads Key team members will also be invited to the kickoff. This often includes the lead programmer, main interface designer, and vendor representatives.

Internal Stakeholders These are the folks around the business who have "skin in the game." They are the internal customers for whom the project is being delivered. This group includes the main business stakeholder.

Company Leadership

In addition to the project leadership team, the company leadership is invited to the project kickoff. This may include the president, CEO, CFO, COO, and possibly even investors or board members. It's important to have company leadership in the room so that everyone is on the same page at the very start.

Who Gives the Kickoff?

Your best speaker should give the kickoff. This is often your project sponsor or program manager/account manager. Another great approach is for different team members to give pieces of the presentation.

Kickoff Presentation

The kickoff is normally done in a PowerPoint presentation, with supporting documentation available such as the scope documents and business requirements. You will cover:

- High-level project definition and project charter
- Business case and metrics
- Project approach and technologies

- Introduction of team members and roles
- Project scope
- Out of scope
- Timeline
- Budget and budget reporting
- Risks, cautions, and disclaimers

High-Level Project Definition

For the kickoff, you'll want to review the project charter, discussed in Chapter 5. Everyone in the room should be familiar with it; but, in reality, people read things and they forget.

Business Case and Metrics

As noted in Chapter 5, your project charter document will have a business case. Now is the time to review it. What savings does the project achieve? What efficiencies? What money will it make? What competitive advantage will be achieved? It's this part of the meeting where you get everyone excited about the purpose of the project. (You'll bring them down to earth later!)

Next comes measurement, as also discussed in Chapter 5. How, exactly, are you going to know if the project has achieved the goals you set out to accomplish? Now is the time to announce how you will measure success. It's important to grapple with success metrics now because you will often need to collect pre-project benchmarks against which you will measure.

Say the purpose of your project is to give the business better data and reporting. In that case, you may need to obtain information from staff about the gaps in the current data and reporting. What problems is it causing? Later, you will report on how the project has filled those gaps. You will go back to the same staff after the project has been completed and collect their thoughts on how the project has impacted their work.

Another common goal of technology projects is to "move the needle." A company might want to gain new customers, acquire new web traffic, or achieve a competitive advantage. In all of those cases you will need to get pre-project benchmarks. Further, you will need to come up with a way to measure that the needle has actually moved.

Why is this important in the project kickoff? Because measurement is tricky. People have a tendency to dispute metrics after the fact. If everyone agrees right at the start how to measure success, it will diffuse conflicts later. Further, as we'll discuss in Chapters 11, 12, and 13, the middle and end of projects are often very painful for everyone involved. Projects are complex, pressure-filled, and rife with compromises. It's good to establish your metrics while everyone still has a clear head and an optimistic outlook.

Keep in mind that some things can be hard to measure. For example, companies get new customers all the time. How will you know if the new customer is related to the project or not? Once again, this is a reason to establish methods ahead of time. In the project kickoff, you might establish that you are going to do a survey of salespeople and get their feedback on whether sales have increased because of the project. This is anecdotal data, to be sure. But if everyone has agreed ahead of time that this is your metric, then you can use it later.

Project Approach

In Chapters 5 and 6 we talked about project approach, research, and technology choice. In the project kickoff meeting you will review these key decisions and the reasons for making them. You will walk through which technologies were under consideration, key discoveries in your research, and rationale for the final choice.

If outside vendors or consultants have been chosen for the project, this may also be a time to review the criteria for those selections.

Team Members and Roles

As we discussed in Chapters 4 and 5, it's very important to get all the project roles covered in some manner and to make sure everyone is clear on their boundaries. By kickoff, everyone needs to put their name badges on, set their toes to the starting line, and be ready to run in their lane.

Share the project roles and responsibilities with the broader group. Let them know who is in charge of what. This helps to communicate who is not responsible for what. You will speak about who can answer which questions and whom to approach with different sorts of issues.

You will also cover the stakeholders in the business and their responsibilities. Make sure to include what you will be asking of them. For example, stakeholders from different divisions of the company may be tapped to do UAT. Mention this now so that people can start to think about who in their unit will be available, weeks or months from now, to participate. It's also good to say that from time to time the business unit stakeholders will be expected to help clarify requirements, make decisions, and review progress. If a question comes up on the project, they must take it back to their unit, collect information, and bring the response to the project team.

Project Scope

In Chapter 6, we discussed project scope and business requirements. By the kickoff meeting, you should have these all finished. In the kickoff, you will review the major project components that will be accomplished in the project. Just as important is to review what is not in the project: out-of-scope elements.

Out-of-Scope

Let's say you are replacing a CRM. The project might include configuring the CRM, a product such as Salesforce, for your company. Such a project would typically include the migration of people's existing contacts into the new CRM. But there might be other items, notes for example, that will not be migrated, or might be migrated in a different format. This is an out-of-scope element and should be noted in the kickoff.

Another example of out-of-scope items might occur in a website rebranding. Many companies have ancillary sites. Maybe there's a separate site for job postings. If rebranding that separate site is not included in the scope, now is the time to make sure everyone is clear.

It's often the case that out-of-scope items are a surprise to audience members in the kickoff. This may be the first time they have heard it. As business requirements get pinned down and the reality of budgets sinks in, features get "de-scoped" without broad communication. This is another useful purpose for the kickoff.

Timeline

You will share the high-level timeline for the project. We discussed timeline construction in Chapter 6. For communication and reporting purposes, you will need a tool that shows not only the timeline, but how you will track progress against that timeline.

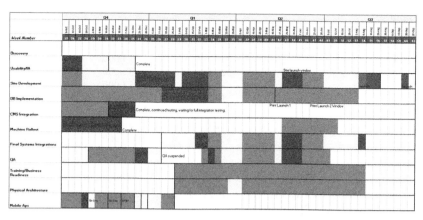

Figure 9.2

A "heat map" is a useful tool for such a purpose. It shows milestone dates and how the project is tracking against those dates. At the kickoff, you will introduce the audience to the timeline and how you intend to report against it.

A typical heatmap can be found at my website (http://www.emediaweb .com/completesoftwarepm) and is reflected in Figure 9.2. As you will see if you download the Excel file, different colors are used to show how a project is tracking. Green for "good," yellow for "at risk" and red for "off schedule/budget."

Of course there are many different tools available to track a project's progress against a proposed plan. We'll discuss those more in Chapter 10.

Often, your program manager will have examples from other projects like the previous one. It can be very helpful to show these examples to the audience as a first introduction of what they will be looking at as the current project unfolds.

Budget

In the kickoff, you will review the high-level budget. This discussion helps to squelch anyone still muttering about things being out of scope. This is also the place where the CFO or comptroller's ears perk up.

For the kickoff, you will be reviewing the budget at a very high level—the major project components and what is estimated for each as well as any additional hard costs, licenses, consulting fees, and contingencies you've planned.

Table 9.1

Item	Budget	Actual	Variance
Planning and documentation	$5,500	$5,000	($500)
Programming	$25,000	$33,000	$8,000
Data migration	$7,500	$7,000	($500)
Design	$4,000	$4,000	$0
Hosting setup	$3,500	$2,500	($1,000)
Licenses	$6,000	$6,000	$0
TOTAL	$51,500	$57,500	$6,000

Then, as with the timeline, you will introduce the audience to your proposed reporting tool where you will be tracking budgeted expenses against actual expenditures (Table 9.1).

Risks, Cautions, and Disclaimers

I often feel this is the most important part of the kickoff.

I always include a list of major project risks in the kickoff. Another useful tool is the trough of FUD graphic you'll find in Chapter 11.

Figure 9.3 may also be shared in kickoffs.

This image reveals that projects start with lots of conflicts and decisions. The darker gray in the figure represents times of conflict and decision-making. The lighter colors represent calmer project periods. Your audience may already be familiar with this! Then, typically, the project unfolds, often pretty quietly. Towards the end of the project is when risks surface, all the trains come together on the tracks, and there can be problems.

Take a look at Chapter 8 on risks and Chapter 12 on problems. I often include examples of these in the kickoff so everyone has a good idea of what may come up. It's a kind of inoculation. Later, when problems do surface, people will say, "I remember you said this could happen in the project kickoff."

Project Beginning   Project Middle   Project End

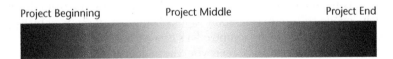

Figure 9.3

Monthly Steering Committee

The steering committee is a project governance body with authority for larger project decisions. A steering committee is formed at the beginning of the project, with the blessing of the company leadership. The purpose of the monthly meeting is to review the health of the project with the steering committee.

Monthly Steering Committee Attendees

The people invited to this meeting will look very similar to those in the kickoff. They include the project leadership, a representative or two from company leadership, plus any other people needed to support the agenda. (For example, a programmer might be included if an explanation is needed regarding the success or failure of a piece of technology.)

While this group is similar to the kickoff, it must be smaller. The project leadership is, of course, in attendance. Company leaders such as the CEO or CFO may be invited, but often don't attend. Typically, they are given a heads-up if their attendance is required because an important decision must be made.

The business stakeholder group is usually smaller than in the kickoff. You will want only the principal stakeholder(s), for example, the most important internal customer(s) for the project:

- Company leadership (CEO, COO, CFO, invited)
- Project leadership (project sponsor, program manager, project manager, tech lead, vendor leads)
- Most important business stakeholder(s)
- Other supporting cast as necessary

Monthly Steering Committee Agenda

The monthly steering committee has a set agenda, which is

- Reporting on timeline
- Reporting on budget
- Key accomplishments in preceding month

- Plans for upcoming month
- Problems, gotchas, opportunities, and discoveries that require steering committee agreement (e.g., budget shift, consumption of contingency, large new feature)

In Chapter 6, we talked about breaking down the project into tracks and timelines. You will select a tool (which can be as simple as Excel or Word) to keep tabs on the timeline. You will also have a tool to track the budget. You'll use all these tools in the monthly steering committee to give the project and company leadership a view into the health of the project.

Next, you'll be like a news reporter, letting folks know the key accomplishments that happened in the prior month. Make sure to spend a little time here and give color commentary. It's the only time many people in the leadership group will have any exposure to the "nitty gritty" of the project. So if a piece of technology is working really well, do a small demo. If programmer Janice worked all weekend to migrate the data because it was dirtier than expected, share this as well.

Then it's time for a look ahead into the coming month and what major elements the project team will be tackling.

Towards the middle and end of the project, you can expect some of your "gauges" to be turning yellow, red, and orange. Problems have cropped up. Perhaps a licensed application simply does not work as advertised. Maybe a team member quit. Maybe a business stakeholder is kicking up a fuss, declaring that some out-of-scope item is simply intolerable. The steering committee meeting is the time to surface these things, propose recommendations and actions, and come to agreement.

For example, if a team member has quit, the only option may be to extend the timeline while a new person is hired. Another option, but usually more expensive, is to retain an external resource to fill in. (In this case, the timeline might also extend, but by less.) The steering committee needs to be informed of the options, weigh them, and decide which is best. It may be that the CEO or company president must also sign off on the decision. If they are not in attendance, the steering committee's recommendation will be brought to them.

The decisions of the steering committee are recorded, and schedules and budgets are adjusted as needed.

Note: Sometimes, a bi-weekly steering committee meeting is preferred rather than once monthly.

Weekly Project Management Meeting

A key purpose of the weekly project management is to make sure the project team is all on the same page with the major tracks of the project.

Weekly Project Management Attendees

The group attending the weekly project management meeting dips deeper into the project team. You will have:

- Project sponsor (invited)
- Program manager
- Project manager
- Vendor project managers
- Programming leads

Weekly Project Management Agenda

The weekly project management meeting is often a very familiar, genial gathering. It usually has a "just us chickens" feel.

This meeting, like the others, also has a set agenda:

- Reporting on all tracks of the project with discussion and problem solving as appropriate
- Identifying to-dos and follow-ups
- Tasks in the upcoming week

You will tick down the project tracks, report on progress, and identify the issues, questions, and barriers that have cropped up. This will generate a list of tasks, to-dos, and follow-ups.

For example, it may be that a business requirement needs to be re-clarified with a business unit. The project manager will note that and schedule a brief meeting. A question may have come up about the contents of an existing data feed. There's another follow-up.

Daily Standup Meeting

The daily standup meeting is a fixture in the Agile methodology. The agenda and attendees are simple. It's the project manager (or "Scrum master" if you are doing Agile whole hog). All programmers report what they are working on, their progress, and if there are any barriers to their work.

The goal of the daily standup meeting is to be brief, very brief. A mere 15 minutes is often sufficient.

Well-Run Meetings

Few things are more important to a well-run project than well-run meetings. There has been a large-scale uprising in the business community against meetings. A big cause of this is that most meetings are so bad.

Following are some ways to achieve well-run meetings.

Insist on Attention

The amount of multitasking going on these days in meetings is ridiculous. Almost everyone is monitoring and answering e-mail while they are supposed to be paying attention to the meeting. A tip-off is when someone is asked to respond, and they say, "I'm sorry. I didn't understand the question. Can you please rephrase?"

And the question was something simple like, "Are you available at 2 pm on Friday?"

Leaders of the project should not allow this kind of partial attention. State up front that the meetings will be well run and purposeful. Tell your team members that you'll keep your end of the bargain if they keep theirs—full attention.

Timeliness

Waiting five or ten minutes for a meeting to start or a conference call organizer to arrive is not okay. It contributes to the tendency to multitask (everyone's doing that waiting for the meeting to begin), and it creates a vicious cycle of lateness. Other team members start to arrive later and later because the meeting never begins on time.

The project management team has control. Start the meeting on time. I have found even chronically late executives will show up on time for meetings if they know this particular meeting starts promptly, is well run, and waits for no one!

Getting "into the Weeds"

In update meetings (e.g., weekly project management meetings), it's important to control too much "in the weeds" discussion. While I am a great fan of "the weeds," the weekly and monthly meetings are not the appropriate time for them.

Another way to express this is "avoid getting into the *how*." Update meetings are not about finding solutions, but about identifying problems and high-level follow-ups that lead to a solution. A phrase you'll hear a lot from the program manager and project manager is, "How about we take that off line for a deeper dive." A separate meeting with a tight group is held exactly for the purpose of getting into as many weeds as possible, discussing all the possible "hows" and proposing solutions.

There is a bit of a balancing act with "getting into the weeds," and experienced project and program managers will be hip to it. We discussed team members' personalities in Chapter 4. The fact is that many programmers' style of communicating *begins* in the weeds. Slowly they work their way up from data elements, data tables, and other minutia to the larger point they are trying to make or question they have. At the end of all this detail, suddenly a big "gotcha" may pop out. A gotcha certainly *is* something that needs discussion in the weekly meeting, and may need to be raised at the next steering committee.

I usually let an "in the weeds" conversation play out just a bit, to see if a "weedy" communication style is masking something important the project team needs to discuss.

Needs to Be Kicked Upstairs

Another thing that gets meetings off track is when the project team spends a lot of time talking about strategic business issues not in their purview. It's the responsibility of the project management team to spot these items.

Say, for example, an external resource is falling behind on their piece of the project. A programmer will report, "I spent seven hours over the weekend working on the mobile app because ACME Technology's widget is causing performance problems." Then the team goes on to discuss how much more time it will take to get ACME's widget working and how another company's widget is better.

All of this work and discussion is off track. It's not up to the programming team to unilaterally decide to make up for a non-performing vendor, nor to select another vendor. The correct response here from the project leadership is, "Wait! Stop! Don't spend any more time! We need to have a business discussion whether or not to proceed with ACME."

Another example might be in an e-commerce implementation. A programmer might report that the cart flow isn't working out as specified in the business requirements, so she needs to implement a different flow. Don't let the project team start talking about the different flows! Not only is this "getting into the weeds," but it is also not appropriate for the team to decide. Changes to something as important as the checkout process need to be brought back to business stakeholders for input.

A good project management team will instantly spot those seemingly innocuous meeting items that need to be "kicked upstairs."

Poor Quality Sound—Speakerphones and Cell Phones

The scene: A bunch of people huddle around a conference-room speakerphone. At the same time, in distant office locations, people are dialing in.

"What? What?" one of the remote-dialers says, "Can Jane move closer to the mic?" Jane moves closer. She speaks. She is heard. Roger speaks. "What? What? Can Roger move closer to the mic?"

Here's your choice: (a) Everyone dials in from individual phones, or (b) Invest in some high-quality acoustics. That means expensive speakerphones and mics throughout the room.

Similarly, a lot of people eschew wirelines these days and only speak on cell phones. Often these connections are poor, headsets are insufficient, and people sound as if they are at the bottom of a well. Compounding this problem is the number of people in noisy office situations or doing conference calls from airports.

It may not be possible to control all of the factors that interfere with crisp voice connections. But too often project leaders do not address the question at all. Simply alerting the team that they should do their best to be in a quiet location with a decent connection for meetings eliminates 70 percent of the poor connection, noisy location nonsense.

Too Much Talk

The leader in meetings usually talks too much. There is so much a project manager wants to communicate! Also, many people love the sound of their own voices. Here's a good rule of thumb: Most human beings can digest about a paragraph of information at a time.

If you are the meeting leader, deliver a paragraph of information. Pause, ask for feedback, or ask a direct question. "Jane, what do you think of this?" In addition to reducing multitasking, this style of communication helps the team absorb the meeting's information. They will not tune out, as so often happens when one person is hogging the floor and won't stop talking.

It is sometimes necessary to deliver a whole bunch of information at once. This might occur, for example, if something new and complex has been discovered and you need to update the whole team. I call this the "frame up" or "context setting."

If this is the case, alert the team you are about to talk a lot. I normally say something like, "It's going to take me about five minutes to describe what we just found out about the database. So everybody hang in there!"

Agenda and Notes

All meetings should have a published agenda and follow-up notes. It's a good idea to have one person regularly tasked with taking the notes and distributing them. A good set of notes records the key points made under each agenda item. The notes end with the follow-up steps and who is on point to address them.

Quality meeting notes are so important I often say, "If there are no notes, then everyone just wasted an hour of their time."

CHAPTER 10

The Project Execution Phase: Diagnosing Project Health; Scope Compromises

s your project moves out of the requirements and specifications stage and into development, you may suddenly feel as if an evacuation order has been issued. Many business folks and project team members may panic at this juncture, because it can feel very eerie when the noise and commotion suddenly evaporate. Don't worry: It's a natural part of being in the middle third of a software development project. If it makes you feel better, behind the scenes your programmers are saying, "*Peace and quiet?!* I am working 60 hours a week and have 200 issue tickets to address!"

What Should Be Going on Behind the Scenes

If a project is going well, business requirements have been collected, in the most highly visual way possible, and the programmers have a clear roadmap of what they need to do. Here are some common activities happening at this stage:

- Programmers are coding, even though wireframes may not be fully done. Many back-end tasks can be done in the absence of visuals. Even many front-end pieces can be roughed in. If your team is working in a more Agile style, bits of functionality are starting to emerge and to be shown to the project team and even business stakeholders.
- Wireframes are being created and signed off. Wireframes, you say? I thought we were supposed to have a ton of visuals already? But remember, the visuals created in Chapter 6 were based on assumptions.

They were, by definition, rough and "the best we could do" under the circumstances. Now, a UX person will refine those ideas, sorting out potential conflicts and issues that might have been glossed over in the attempt to sketch out a broad picture.

- As the UX person creates wireframes, a designer will be doing the first visuals, demonstrating how those wireframes can be enhanced to create a beautiful piece of software.
- The project manager is tracking everyone's progress and seeing where things are going off schedule. At this state, he or she is inevitably reporting, "Fair weather! Clear sailing."
- Initial testing plans are being considered.
- The business analyst is clarifying specifications. Remember how, as we collected business requirements, our goal was to be visual? Now we're getting down to the nitty gritty. A programmer may have a question about what something meant. People may say something like, "Wait, I thought I understood that, but can we go over it again?" We covered this in our Chapter 7 about budgeting. Sometimes the obstacles that surface here may require a long discussion to resolve (more about that in a minute).

One common reaction at this stage is to think, "Wow! This is great! I'm so glad things are going so smoothly." But the truth is, things *should* feel like they're going smoothly at this stage. They *always* feel as if they're going smoothly, even in software projects that are ultimately doomed. For example, if too many risks have been left on the table or poorly addressed, they will be hiding right now, waiting to ambush you later. If you have an unskilled software team, pretty much no one will be aware of it yet. Unfortunately, poor code is usually only evident when you are quite close to launch.

Seasoned software teams who consciously accepted defined risks at the start are bracing themselves during this quiet time, preparing strategies for dealing with risk consequences later.

The Best Thing You Can Ever Hear: "Wait. *What* Was It Supposed to Do?"

In the midst of quiet, sometimes further discussions of business requirements are necessary. People can get very impatient with this phenomenon, and chastise team members or programmers saying, "We already talked about that. I can't believe you don't have that straight yet!"

To the contrary, if this is happening, celebrate! It is a very healthy sign—one of the best you can see at this stage of a project.

The truth is, no matter how much visualizing and documenting you do, features and functions will still not be understood in detail. I am always surprised how, even with a good visual, two people can look at the same thing and have utterly divergent understandings. My experience has been that human beings absorb information in a layered way. An initial grasp of a concept is followed by a more in-depth understanding, which is then followed by a real internalization of the concept. In short, business requirements need to be understood, re-understood, re-re-understood and even re-re-re-understood.

A common mistake project leaders make at this stage is to frown on a team that goes back into discovery and business requirements mode. "We are supposed to have finished discovery! With all the documentation you have, you should be clear on the requirements!" It can feel like the team is taking a step back and wasting time. Stakeholders might begin to feel that the project will never get done.

Do not fall into this trap. In the real world, discovery takes place up until the very last day of your project. We might wish it could be different, but it's not. Encourage these conversations, even though it may feel stressful to hear that what you thought was well understood is actually not.

I feel so strongly about this that if I don't hear discussions going on at this stage, I'll poke the hornets' nest myself. I might say something like, "Are we sure we're all on the same page with that admin module? Can you explain to me how it's supposed to work?" Someone will inevitably give a shaky answer, which allows the group an opportunity to re-re-re-understand.

Neutral Corners

Another characteristic of this stage is that people may seem as if they are working in isolation. In your head, you imagined collaborative programming teams all huddled around a workspace solving problems together.

In the real world, this phase of a project generally involves everyone's going to neutral corners and working in half or even complete isolation on their particular tasks. Later on in the project, the work streams will come together.

What If Things Aren't Quiet?

If things are not quiet (aside from clarifying requirements), you've got trouble. Commotion is not a common phenomenon at this point. Even in software projects destined for real problems, the working phase will feel quiet.

One problem that can crop up in this phase is disagreements among programmers. Typically, this manifests itself in one programmer saying the other programmer is unskilled, uncooperative, or doing something the wrong way. If that starts to happen, it's time to pull the emergency brake and bring the project to a temporary halt while the conflict is addressed.

Programming teams are like army platoons: A high degree of trust in one another's competency is absolutely critical to the success of the campaign. If one guy thinks the other guy doesn't have his back, the platoon will never take the hill.

I have found that it is very hard to know which engineer has the correct understanding and which is the one lacking in skills. About half the time, the one who blew the whistle ends up being the problem team member, but it's just as likely to be the other guy or gal. If you have a trusted senior programmer, he must make the call. But, to be frank, a team with a trusted senior programming lead will rarely run into this issue; a seasoned programmer will insist on working with proven team members from the beginning.

Sometimes it happens that your trusted senior guy has suspicions about a vendor that has been hired. If he does, listen to him. It's possible that his suspicions are unfounded, but it actually doesn't matter. Trust and collaboration are the key elements: Without them, you're doomed. The question of who's right and who's wrong matters far less than you probably think.

In situations where the programming team is unfamiliar to you, pay attention! I have found this dispute phase is the first sign of a bad/unskilled/unproven technology team and is potentially fatal to the project. You may need to dismiss the whole team. A good team will feel calm and serious; they will be very cooperative with one another, rarely letting the situation devolve into intense disputes that require project leadership intervention.

Making Decisions

In this phase, however, smaller decisions are getting made all along the way much of the time without business stakeholder involvement. This is right and correct. It's also why things may feel very quiet now.

However, some decisions will inevitably land on project leadership in this stage. They will usually involve a strategy or approach to a programming task. The tech team will become aware of something—perhaps a decision needs to be made that will impact the schedule or budget, and the issue bubbles up the project leadership to be taken to the steering committee.

How to Listen to the Programmers

When an engineer or project manager begins to talk about an issue at this stage, it may feel as if she's dragging you *very* far down into the weeds. To a business stakeholder's ears, it will all sound like, *"Blah blah* MySQL, *blah blah*, data migration, *blah blah*, legacy system."* It may take all the patience a businessperson has to keep listening and repress the urge to throw the programmer out of the office.

Have faith that somewhere beneath all the *"blah blah blah"* there might be a key issue needing attention. The engineer's communication style is to start with the details, in a sort of an inverse pyramid. This is typical of programmers and of tech people in general, including project managers and BAs. You need to let the tech person move through the details and work up to the big picture, then you may need to go back and ask her to repeat the details. Often, it is not possible to get her to state the big-picture problem up front and give supporting information.

What kind of decision will you be facing? It really depends on your project. But now we'll cover some common themes that are usually at play.

The Programmer's Prejudice

Programmers have two very strong biases to be aware of. Programmers believe:

- If a task is boring or laborious, you as a programmer should write code to deal with that task.
- If, at the end of a software project, tasks that could be automated are still being done "by hand," you as a programmer have failed.

These biases make sense: After all, computers were invented to automate human tasks that seem boring or laborious. However, such biases can be dangerous to your project. Be on the lookout for them in your programming team, because they can influence key decisions in a way that is not healthy for a project.

A programmer might say to you: "We need to decide which direction to go. If we don't do this additional programming, people will still have to perform the important task *by hand.*" Or he might say, "In order to transfer all that data over to the new system, I will need to write ten scripts to clean the data and import it. Otherwise, someone will have to comb through all the data *by hand*." The phrase "by hand" is wielded like an expletive, and the way engineers say it can be enough to scare you to death.

Remember the piles of snow. Be on the alert for people trying to minimize piles of snow by writing custom code. Ask yourself: What is so bad about having Robert take a week to go through the spreadsheet and make sure all the zip codes are in? Sure, a programmer *could* write a script to normalize the data, but is it necessary? A programmer's time is usually considered the most valuable in a software project. Can't we just have Robert shovel that pile of snow? After all, what if the simple just-write-a-script task ends up taking five weeks?

Your programmers may fight you on this. Remember, they were trained to automate laborious tasks and to leave no by-hand steps. Decide what automation is appropriate and what piles of snow are better to shovel. Listen to them, but make your own judgment while being aware of their biases.

SneakerNet and the Fred Operating System

"SneakerNet" and the "Fred Operating System" are terms I use in dealing with the common decisions cropping up in the implementation stage of a project.

SneakerNet Integrations

In Chapter 8 we discussed the risks inherent in integration points—when one system needs to exchange data with another system. Programmers will strive to automate all integration points. Again, this stems from their desire to eliminate any "by hand" steps, especially those involving data.

SneakerNet is a term to describe a by-hand integration point. In order for the data to get from one system to another, Betty needs to save the data on a flash drive, get up from her desk, and walk over to accounting—using her sneakers—and deliver the data. These days, SneakerNet is often more like Betty exports the data into Excel once a week and e-mails it to Accounting. A programmer might strongly wish to automate this phase, but there is nothing inherently "wrong" in choosing a more manual strategy. Selectively and consciously choosing some SneakerNet steps can significantly reduce the risk in a software project.

Your programmers will want to avoid all SneakerNet integrations. However, at the implementation phase, new integration points will be discovered and difficulties with known ones will crop up. Seriously consider the option of leaving a few SneakerNet points in place—at least temporarily—rather than trying to solve all your integration issues at once.

The Fred Operating System

When you reach the implementation phase, it may turn out that some feature or function is far more difficult and costly to program than originally thought. Perhaps a tool wasn't all it was advertised to be, or, once the business requirements are re-re-re-understood, it's found that the task to program the feature was far bigger than originally estimated. What ends up on the desk of the project leadership is a request to authorize more programming time and budget because this discovery has been made.

But Fred knows how to do the task! In fact, it's been part of Fred's job all along. It's the Fred Operating System.

Similar to SneakerNet, there is nothing inherently wrong with the Fred Operating System, though your programmers, in describing the situation to you, may make it sound like the worst thing in the world. Again, their biases are at work. Sometimes, the best thing to do is to leave the Fred Operating System in place, at least temporarily. Perhaps put some automated tools in Fred's hands to make his job a bit easier.

The Hidden Benefits

You may already be suspecting that the FredOS and SneakerNet are, at their base, a way to phase a project, making it smaller and, by definition, reducing risk. This is true. Leaving in some human factors—by-hand work and integrations—can be a wise thing to do along the way.

Furthermore, by living with your temporary SneakerNet and Fred OS, you may make discoveries that ultimately inform and enhance your final, all-automated solution.

Demos and Iterative Deliverables

In Chapter 6, I advised making business requirements as visual as possible. The idea is to allow business stakeholders to actually *see* rather than just read about what the software will do.

This visualization process needs to continue throughout the development of the software. As a practical matter, this means bringing business stakeholders into the process as frequently as possible so they can see and give feedback on pieces of the software as it gets built.

Why Iterative Deliverables Are Important

Even if you have highly visual business requirements, you probably do not have a piece of "clickable" software. Some very wise enterprises invest in a smaller proof of concept (POC) phase. A proof of concept is a small implementation of the software showing key features and functions actually working. If you do not have a POC, the business stakeholders have only seen mockups. Even with good mockups or sample screens, nothing, absolutely nothing, replaces the live experience of clicking, whether on a desktop or mobile device. You can see it in people's faces. As they click on a live piece of software, you can watch the dawning realization, "Oh! I see! This is how it works." Even very seasoned software developers, myself included, have this reaction.

The most important thing the Agile method has taught us is that people must click on something real in order to understand and give feedback.

Seeing mockups is good. Clicking is better.

Why Iterative Deliverables Are Hard

Seems as if it should be easy, right? "Sure, when such and such a module gets done, we'll just show it to the business stakeholder." It's true that in some projects, it works exactly like that. Simple, small- to medium-sized web projects are an example. As certain pages or page features are completed, you can just show them to businesspeople and get feedback.

In other cases, it's much harder. The first point of resistance may be the programming and engineering staff. This is especially common if programmers have had a very "waterfall" background. Remember from Chapter 2, a Waterfall style means the programmers take business requirements and "go off to a cave" to program, then deliver a final product.

If an engineer has used this working style all her life, it may be very difficult to switch. She may feel interrupted by showing pieces to the end customer. She may like the quiet weeks of programming before having to return to the chaotic back-and-forth process of receiving feedback. Moreover, if someone sees a piece of functionality, they might have input. Or, worse, *change their minds*. Remember when I said the things that programmers hate most are shifting business requirements (more about how to deal with that later).

The second challenge in showing pieces of software has to do with the business stakeholders. There is a wide variety in people's ability to see a partially done piece of software and envision the whole. Some people are very good at it. You hear them saying, "Oh, I get it. I click here and this happens. And I can see when this piece is done, then this other thing will happen."

Other folks just can't seem to build this kind of picture in their mind. In this case, you will see frustration. One click will be successful and produce that good "dawning realization" reaction. The next click might be blocked because that piece of functionality is not complete. Then you will see frustration. The business stakeholder might say something like, "I don't get it.

Nothing happens when I click here." A good project team will keep reminding the business stakeholder what they are seeing is partial and iterative, but sometimes it can still be difficult. I have seen reactions range from frustration, bewilderment, and disengagement at this stage.

It's particularly problematic for certain business stakeholders if the design is not in place. Someone who places high value on the final design may not be able to get his head around the functional clicks without a full sense of the visuals.

The last challenge to iterative deliverables happens at the level of the software itself. Some projects, or aspects of projects, are what I call "anti-iterative" or "anti-Agile." In other words, it's just plain hard for the programming team to deliver bits of functionality to be reviewed.

Take the case of a software project involving multiple integrations. To a great extent, these integrations may need to be in place for anyone to see anything. Data migrations can have a similar impact. All or much of the data needs to be in the system before people can see how the software works. Or, it might be in a large software project that different teams have to be done with their different tasks for anything to be "clickable." E-commerce implementations are particularly hard to show iteratively. Until the software is done and hooked up to the payment gateway taking real credit cards, no one can get a true experience of what the payment process will be like.

What You Can Do to Achieve Iterative Deliverables Even if It's Hard

You can still achieve iterative deliverables even if you find yourself in one of the previous situations.

Someone once gave me this piece of advice: "The responsibility to be understood rests with the communicator, not the communicate-ee."

Put another way, it is the responsibility of the project team that the business stakeholders "get it." If the business stakeholders are not getting it for whatever reason, you have to make sure that they do.

This piece of advice often produces resentment on project teams. This is understandable. Especially if the business stakeholders appear that they aren't trying to comprehend or seem unusually dense (sorry, business stakeholders).

Remember, the business stakeholders are your customers or proxies for your customers. If the customer ultimately rejects the product, the project will have failed. It's best if the project team simply accepts this reality up front and develops strategies to somehow get something clickable and understandable in front of the business stakeholders.

Here's how to do it.

Demos

Demos are not the real software. A demo is a kind of "straw man" implementation allowing a business stakeholder to see more, even if the actual software is not fully built.

Often, I have a programmer, usually a front-end guy, look at what the project team is building and then create a separate demo. He knows what the software is supposed to do. So he creates a "fake" that does it.

You will encounter all kinds of objections here. Waste of resources for one. Another thing you'll hear: "What they're seeing is not real. Why can't they just look at the piece of functionality in the development environment?"

All of these objections have their merit. But in 20 years of software development, I have never seen demo time wasted. Remember that programmer from Chapter 6 who said, "Anna, I hate when you make us do this. It seems like such a waste of time, but it pays off in the end."

Demos solve the design problem. If it's not possible for the programming team to apply the design yet in the actual development environment, the straw man team can apply it to the demo. Finally, if integrations or data migrations are blocking iterative deliverables, the straw man team can use sample data sets or simply hard code the demo so that it appears to be drawing from actual data.

Such straw men must be created in cases where a company has to do a tradeshow or some other kind of presentation.

If you're saying, "This sounds like a proof of concept," you are right. If you didn't do a POC to begin with, inserting demos as a kind of "mini POC" into the middle of a development project is a way of achieving the same result.

A reminder: The result you are after is to make sure the project is tracking with the business stakeholders' expectations. Static visuals, while a good start, are not enough. Nothing replaces clicking.

Scope Creep

As hinted earlier, the purpose of producing iterative deliverables is to get feedback and make sure the project is tracking with expectations. Ironically, getting feedback may be what the project team fears most. In software terminology, it's called "scope creep." Scope creep is comprised of the bits of feedback and questions like "can we just add this?" and "can we just change this?" One by one such items of feedback add up and create a larger project, in time, money, and scope.

Here's a skeleton in the closet of software development: Teams often don't like feedback. Frequently, they would prefer to be left alone and do the software. Buried in this preference is a sense, often unconscious, that if the software is delivered fully baked, it will be too late for the business stakeholders to change their minds. It's a way of resisting scope creep.

Dealing with Scope Creep; Early Is Better

The problem with the skeleton-in-the-closet is that the software developers are right. Once the software is done, it *is* too late for most feedback.

Imagine you were making a soufflé. Once it comes out of the oven, it's way too late to decide you want your vanilla soufflé to actually be chocolate—even though this *sounds* so simple in concept. Just get that chocolate flavoring in there. But when the soufflé is baked, this simple concept becomes impossible.

So it is with software development. Unraveling layer after layer of programming can be as hard as un-baking a soufflé. The earlier you spot changes, the more possible they are. And the cheaper they are.

Thus, the earlier you can provide a clickable look-see to the business stakeholders, the better everyone's life will be in the end. This is another reason straw man demos can be useful. It may be the only way in an early stage of a project to give business stakeholders a sense of the software.

Scope Creep and Budgeting

The budgeting method presented in Chapter 7 allows for a certain amount of scope modification. The percentage assigned to "additional discovery" is another way of assigning a portion of the budget for shifts and changes. A feature is understood at a high or medium level, but still more discussion needs to take place. If the additional discovery can be accompanied by something clickable, "Here, is this what you were envisioning?" then you are really on the right track.

Further, the funds set aside for contingencies help in dealing with scope changes.

Scope Creep and Governance

As I have noted again and again in this book, when you are creating something innovative, it is often impossible to know how it's going to work out. We've talked a lot about technical feasibility, such as whether the project team can actually achieve the innovative business requirement. Now we are talking about scope change, or, simply put, the business stakeholders changing their minds.

Here is a real-world example: A project team is working on a patient portal in the veterinary industry. This portal has been requested by pet owners who want to be able to log in and see their pets' test results. The programming team creates the portal and puts the initial functionality in front of some business stakeholders, the vets themselves.

Immediately a flurry of feedback comes in. It's great! But now that we see it, we suddenly realize the need for an e-mail component so pet owners can be alerted when there are new files for them to retrieve. And while we're at it, can the e-mail link right into the actual new stuff?

At this juncture, everyone might start finger-pointing. It seems so obvious. Why didn't anyone think of this before?

My advice: Stop the finger-pointing. Even with seasoned software teams and stakeholders experienced in software development, this kind of thing happens. It's better to accept the fact: Nothing replaces clicking. It's where you'll spot missed requirements and get truly useful feedback leading to the best product in the end.

In the previous example, the project leadership will know that we've gone beyond the kind of back-and-forth clarification embedded in our "additional discovery" budget line. The business stakeholders are adding several new features to our patient portal. It's time to tap the contingency.

The budgeting advice in Chapter 8, you may remember, suggested setting aside some funds in proportion to the amount of innovation, or "creation," in your project. And the scenario described earlier explains why. Especially when creating something new, neither business stakeholders nor the programming team can fully anticipate feedback and reaction. There is a greater probability for scope expansion.

This is where the project governance comes in. The project steering committee would be brought in to decide whether contingency funds should be released to cover the expansion in scope. The steering committee might also be able to cut other features, especially if little or no contingency exists. In the earlier example, it would seem improvements to the patient portal are well worth some additional investment or feature trade-off.

Types of Scope Creep

I often say scope creep always falls into one of three categories:

1. Whining
2. Misunderstanding
3. New opportunities

It's up to the project governance group, the steering committee, to differentiate among these if time and money are impacted.

"Whining," in my book, has to do with fussy, irritable, and hard-to-please business stakeholders. Sometimes whining can come from a business stakeholder whose preferences did not get priority in the project. It's up to the steering committee to make the call and tell the fussy business stakeholder to pipe down.

Misunderstanding is natural. Remember I said you would be understanding, re-understanding, and re-re-re-understanding. Sometimes you repeat this about six times before a project team and business stakeholder group finally break through. In fact, a feature may be all but complete before the *ah-ha* moment. Hint: Misunderstandings of this degree should represent a small percentage of the project. But they do happen. It's very normal. Sometimes, despite the small percentage, such misunderstandings consume a lot of "air time" in the business. Business stakeholders worry that if this feature was so deeply misunderstood, are others as well?

It's useful to control panicked reactions at this point. Misunderstandings about how a feature or two will work are common and generally do not indicate a project going off the rails.

"New opportunities" were spotted in the vet software example. Such discoveries are also normal and to be celebrated. Some uncomfortable choices may need to be made about cutting other features or expanding budget. But, remember, innovation, business opportunity, and competitive advantage are major reasons we undertake software projects to begin with.

Scope Creep and the Team

I've suggested in this chapter that the team may struggle with feedback from business stakeholders. It's understandable. They know there is a timeline and a budget! They are on track to meet it! If only the business stakeholders would quit it with all the feedback. Besides, why didn't they just make up their minds in the first place and deliver a definitive list of requirements?

It's important to acknowledge the truth of what the software team is saying when you hear these or similar statements. Change is hard, and can be frustrating. That said, it's even more important to shift the team out of this line of thinking.

In order to be "agile enough," the development team needs to be flexible. This expectation for flexibility needs to be established from the get-go. Some software teams with Waterfall backgrounds have been led to believe that it is possible to get an absolutely thorough list of requirements up front and that it's the business stakeholders' responsibility to get those requirements. This is a fantasy and the software team must be disabused of it. You can spot this belief when programmers keep insisting on business requirements. Figure 10.1 can help.

Figure 10.1

In reality, business stakeholders and project team members collaborate to arrive at final requirements, and the process continues throughout the life cycle of a project. Business stakeholders do not have some "magic list" in their heads they are failing to disclose. In many cases, they need to work with the programmers to understand what's possible. Yet, many a project team has a belief that it's appropriate to blame business stakeholders for unclear requirements, misunderstandings, and mind changing. A method of enforcing blame is the dreaded change request form. Business stakeholders, like misbehaving children, are sent away to document their change request, and the software team estimates how much it is going to "cost" them—the penalty.

Needless to say, this us-versus-them mentality is not helpful in achieving a successful project. Therefore, it's critical to create the expectation for developers and others on the implementation side that they will need to be flexible. Project team members should be exposed to the methods of collecting requirements, budgeting, contingencies, and governance. This way, they can understand a context has been created to permit scope changes, but also to keep them from getting out of hand. The team will be able to see that changes in budget and timeline are the responsibility of the steering committee.

CHAPTER 11

First Deliverables: Testing, QA, and Project Health Continued

L et's take a moment to review where we've been.

The Project's First Third

The first third of the project was occupied with preparation, which involves:

- Roles, team building, and conflict resolution
- Discovery
- Risk identification
- Business requirements

When reaching the end of the first third, we had enough understanding to budget the project and a reasonable amount of confidence that there was a shared understanding about and agreement on what we were going to program.

The Second Third

Then came quiet, discussed in the previous chapter. This is the second third of your project. Underneath the quiet there is a lot of re-re-re-understanding going on and negotiation on scope changes. Now we're just about to see things come off the production line. So our timeline looks like Figure 11.1.

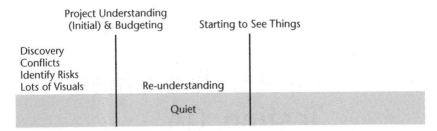

Figure 11.1

A First Real Look at the Software

As we enter the final third of the project, enough coding has been done that businesspeople see the first true "major deliverable." What do I mean by major deliverable? Even if you have done a lot of mini-deliverables and demos along the way, at some point, all the pieces must be put together. Traditionally this is called the "alpha stage" of software development: the first true look-see at all of it, or most of it. If you have not chosen to or been able to provide a lot of intermediate check-in points, this will be the very first time the business stakeholders get a look at the software.

What happens now?

Panic ensues.

Why? Because software at this initial stage is very messy. Things can look broken and wonky and incomplete. After weeks, perhaps months, of silence and quiet working, *this* is all there is? What was everyone doing? It's not supposed to look like this, right?

Actually it *is* supposed to look like this. Even when a project has a good team, sound code, and diligent work all around, the first time you see things all together, they will often be messy and broken. It's perfectly normal.

HINT: KEEP CALM AND CARRY ON

Endeavor to control any horrified reactions at this stage. If you react badly, your technical staff will learn not to show you what they are working on. And it's very helpful to see things as early as you can.

Guided walk-throughs are the order of the day. A project manager or program manager should shepherd the key business stakeholders through the product to confirm features and functions. Resist the urge to instantly point out every single thing that does not look right. It's not just useless, it's worse than useless. It's a distraction and can be destructive (see the previous hint about deterring people from showing you things). The technical staff knows things are messy. You need to focus on asking questions and providing feedback the tech team needs.

You can help by asking *what specifically* you should be giving feedback about. A very useful statement is something like, "Well, I can see that the visual layer is still in process ... and the registration module as well. What are we looking at today in terms of feedback?" This sort of approach allows discussions at this first-look stage to be productive, without you feeling required to critique aspects of the project that the team already knows aren't fully baked. A question like this will also highlight aspects that the team thinks are further along than they actually are. For example, the team might respond with, "Wow, we thought the visual layer was really close. Why do you say it's still 'in process'?"

A word about vocabulary: As noted earlier, those familiar with the formal stages of software development would call this first look the "alpha" stage. When things are in a stable, testable stage, the term "beta" is used. If these terms are meaningful to you, great! I have noticed that these traditional terms are less frequently employed these days, and, therefore, I am not using them frequently.

The Trough of FUD

The general panic and disappointment that arises when a normal businessperson looks at a software product in the first-look stage can be characterized as the "trough of FUD." FUD stands for Fear, Uncertainty, and Doubt.

The Gartner Group has a now-famous tool called the Hype Cycle, which graphs the elevated and then deflated expectations that happen with regard to technology. (For more detail, see Gartner's explanation of a Hype Cycle at http://www.gartner.com/technology/research/methodologies/hype-cycle.jsp.)

The Hype Cycle curve can be applied to everything from the investment patterns around technology start-ups, to social media hype, to emerging technologies. I have developed a similar tool that tracks the emotion and dreams of a software development project against the objective reality of what it was destined to be all along (Figure 11.2).

As we discussed in early chapters, there is a fair amount of excitement and euphoria at the beginning of a software development project. When the project reaches the first-look stage, for all the reasons outlined earlier,

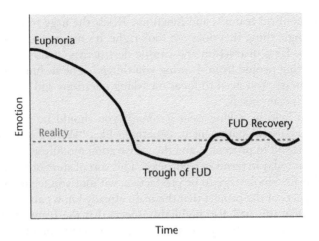

Figure 11.2

business stakeholders become disillusioned. The software looks messy and broken, which is an uncomfortable feeling. But another factor is also at play: This may be when people are first coming to grips with what the software will actually do in the real world, instead of in their imaginations. If frequent demos or iterative pieces have not been provided along the way, this feeling is particularly strong.

In Chapter 6, I referred to an analogy about butterflies. When an idea is just an idea, it is light and airy and colorful, like a swarm of butterflies. When it comes into a real-life manifestation, it's like catching the butterflies and pinning them to cork. And so it is with your software project. While it's still in the idea stage, the project could be anything. If you're one of the lucky folks following the advice in this book, with all your visuals, demos, and iterative deliverables, you and the business stakeholders around you should have a better-than-average idea of what the software is actually going to look and feel like. Still, nothing replaces sitting in front of a cohesive piece of software. This is when reality—and frequently disappointment—sets in.

Notice the light gray line on the graph? This is "reality." It stands in contrast to both the business's initial dreams about the software and their disappointment about it later. The black line represents the emotions of the business, overoptimism at first, then extreme disappointment. In short, the black line is perception; the gray line is reality. You'll notice the trough of FUD. This is where the perception of the software dips way below "reality" into unrealistically negative impressions. Then, gradually, things level out—the business stakeholders' impressions of the software come to merge with what's concrete.

Distinguishing a Good Mess from a Bad Mess

There's messy, and then there's *messy*. Sometimes, at the first-look stage, the particular way in which things are messy just "doesn't smell right." A seasoned software professional (remember your technology partner) will usually be able to spot it. He or she will say something like, "I know things normally aren't pretty at this stage, but this really bothers me."

Commonly, it takes someone with project experience and a tech background to separate the things that are "normally messy" from the things that really shouldn't be. But there are a few rules of thumb. You should be concerned about messiness if:

- What's messy and broken is the straightforward stuff (e.g., the piles of snow and Ikea desks discussed in Chapter 1).
- The workflows aren't in place or don't make sense. (The term "workflows" describes the paths through the product.)

Sometimes technical staff will say that they have been working on all the "hard stuff" (such as integrations or custom code), and they will get to the straightforward stuff later. There is some logic to this approach, but I don't like it. At this stage, it's far preferable to have the Ikea desk/pile of snow tasks well under way. And user navigation from one step to another (workflows) should *definitely* be clear, even if there are breaks in the process.

It should also feel as if the technical staff is grappling with the harder challenges. Approaches have been thoroughly discussed and coding is under way. Folks may have gone through their first theory of how something should be done, and may be on to their second or even third. It also often occurs at this stage that you get your first hints where some aspect—be it an integration or a piece of new technology—looks as if it is not going to work out and Plan Bs are put on the table.

An exception here is data migration, which often must occur at the end.

An Important Checkpoint

I want to take a minute to emphasize what an important checkpoint the first-complete-look stage is. Until now, you've largely been on the dark side of the moon. In other words, your project is like a spaceship that has disappeared from view for a while before you are able to see it again. By following best practices (visuals, iterative deliverables, and demos), you will have a higher-than-average likelihood that the team understands the requirements and is executing a mutually agreed-upon plan. That said, until you reach the first-look stage, you can't really know how all the pieces are going to come together. This is absolutely true in medium- to large-sized projects.

No one knows; not even your senior technical partner. Now is your first chance to validate if the programmers and technical team are going to pull it off. This is a key moment because you're still at the point where you can recover from some problems if they crop up. Soon, it will be too late. It pays to devote a lot of attention at the first-look stage: Figure out if you have a mess or a *mess*, and evaluate where your various risk factors stand.

Finally, the first-look stage is the first real chance you have to reckon with where you are against your original timeline. Now that you have seen the software coming together, you can have a much better grip on how far you have to go.

Getting to Stability

Gradually, through the weeks that follow the first release, things will start to shape up, as that lumpy dough you put in the oven begins to rise into the golden loaf you expected. All the Ikea desks get assembled, the piles of snow get shoveled, and the heart surgery patients come out of the operating room and are in recovery. (Note: Chapter 12 will discuss what to do when none of this is happening.)

First Testing and the Happy Path

In software development and testing, the "happy path" refers to the expected, normal usage of the software. For example, in an e-commerce site, a happy path might involve browsing for an item, finding it, putting it in the cart, browsing for another item, putting it in the cart, and then checking out. The opposite of the happy path might look something like this: Search for an item, put it in the cart, remove it from the cart, put it in again, close the browser, open it again, go to the cart to buy the item, buy it, decide you don't want it, and cancel the order.

After you have had your first look at the software, you and your team will want to focus on making sure all the happy paths work. Once they do, you have reached your next real milestone. As noted earlier, formal software jargon would call this a beta. What this really means is a stable product: The happy paths work, and you are ready for more comprehensive testing: full-on quality assurance.

Quality Assurance

Quality assurance (QA) is the most overlooked phase of software development. Business stakeholders frequently underestimate the importance of QA, assuming that their own people can handle it. No one wants to spend the money for QA; after all, it's just running through the software, right? We can get everyone in the company to do that! But the sentence, "The internal team will do QA," can be horrifying to technology professionals. I can assure you, it's horrifying to me! And it is not appropriate on a medium- to large-sized software project.

To understand why, let's discuss what QA involves, then talk about what levels of QA are appropriate. QA begins with defining what are called "use cases." Think of the happy path example. Use cases are all the journeys through your software that a potential user could take, both happy and "unhappy." Someone needs to sit down and map out all the possible journeys so they can be tested. This sounds like a daunting task because it is. It's mostly a pile of snow–type of task, but it's a big one.

Many of the paths and items that must be tested are those that might not occur to the "normal business user." One is the "close the browser and open it again" scenario from the previous e-commerce example. Another test might be, what happens if two people log on simultaneously with the same password? Other use cases have to do with security-related issues, like testing the input forms on a site to see if they are vulnerable to hacker exploitation. The QA process is the systematic run-through of these use cases in the software.

Is your internal team capable of coming up with all those paths?

Bug Reporting

Bug reporting is the data that comes out of running all these use cases. When something does not work as expected or completely fails, a bug is reported.

Bug reporting is usually handled via a bug tracking system, such as products like Jira or Bugzilla. Yes, you can track bugs on spreadsheets and Word documents. But this gets extremely messy and hard to manage, as different versions of the documents start to proliferate, and no one is sure who is in possession of the definitive list.

Bug reporting itself must be done in a disciplined and consistent way:

- Summarize the bug in a brief description.
- Attach screen captures, often several, of the user behavior that produced the bug.
- List the steps required to reproduce the bug.

In a bug-tracking system such as the ones just noted, the bug is then followed through the stages of its "life." It is assigned to a programmer as "open." After the programmer has addressed the problem, she will then mark it as "fixed." Or, if the screen captures, summary, or reproduction steps are not clear, the programmer may respond with some questions. Once the bug has been fixed, another team member confirms the fix before the bug is marked "resolved." In a final stage, the bug is marked "closed."

Regression Testing

As one bug is fixed, it might introduce another bug. Going back to re-test the software to make sure the fixing of one bug has not broken something else is called "regression testing."

Bugs: Too Many, Too Few

I am often asked, "How many bugs are too many?" Obviously, this is a hard question to answer. For most medium-sized software products that undergo a good testing effort, you can expect bug lists in the hundreds. Bug lists in the thousands or above usually indicate a big problem, unless the project is very, very large.

Interestingly, I never get asked, "How many bugs are too few?" But too few bugs is common in in-house testing endeavors, where not enough test cases have been generated and the testing effort is very superficial in nature. My general, very unscientific rule of thumb is that fewer than 100 bugs means the testing effort has not been sufficient.

Testing: The Right Amount for the Job

When is it okay to have an internal testing team? The answer, once again, comes down to risk.

First, you must consider the technical aspects involved, such as integrations and data migrations. Some items are very hard for civilians to test thoroughly because they may not know how to determine if integrations are really working or if the data is "clean." If your project involves a lot of these technical elements, you would be well advised to consider hiring an outside QA firm for your testing effort.

The other area to consider is the potential for bad end-user reaction. What is the consequence if the software is launched with bugs? For example, you might ask yourself:

- Is the product being used by important customers? Will bugs be embarrassing or even damaging to your image or relationship with those customers?
- Is the "brand" image at stake?
- Is this a transactional system where actual dollars are at stake if the software doesn't function correctly?
- Is the data in the software critical to company operations?

If the answer is yes to any of these questions, you would be well advised to consider a formal, professional testing effort.

If we accept the fact that all software has bugs, then it's really a choice about who debugs it. Rest assured: If a good QA process does not do it, your end users will.

Too Much Testing?

As you might imagine, too much testing happens much less frequently than too little, but it does happen. QA firms will test a piece of software within an inch of its life if you let them. That's what they do.

If you are getting the feeling that your QA effort has been going on forever, it's time to evaluate where you stand. Perhaps there is still some cleanup to do on the user interface. Perhaps the software is still buggy in certain versions of Internet Explorer or doesn't run on an older iPhone, but all transactions and data manipulations are clean and solid. That's the time to think about ramping down your QA effort.

Bug Cleanup Period

If you are working with an outside software firm, that company will offer some period of time in which they will fix bugs "for free." This window usually ranges from 30 to 90 days. Obviously, a firm can't continue on debugging the software forever. This is another reason it pays to do a full test prior to launch.

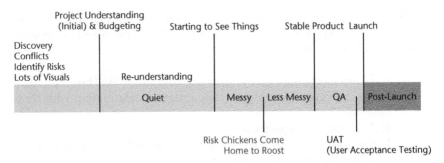

Figure 11.3

Timeline So Far

Let's take another look at the timeline so far and into the future (Figure 11.3).

We have covered software development all the way through QA. In Chapter 13, we'll look at user acceptance testing, launch, and post-launch.

But right now, we're going to take a step back and look at problems—what happens when the software doesn't shape up and the "risk chickens" come home to roost.

CHAPTER 12

Problems: Identifying and Troubleshooting the Three Most Serious Project Problems; Criteria for Cancellation

I recognize that many readers may pick up this book and flip right to this chapter because they are two-thirds of the way through a software project and have problems. In fact, you may be the unlucky person who was asked to step into a project at the two-thirds mark *because* of all the problems. If that is the case, I encourage you to refer also to Chapters 5 to 11 for the background necessary for understanding what you're facing.

As discussed in the previous chapter, there is a "dark side of the moon" phenomenon in software development, especially one that isn't "agile enough." Business requirements get settled and then the programming team goes off and develops. At the first-look stage the business finally gets a taste of the software. And that point—about the two-thirds mark—is when things can come crashing down.

A Rule About Problems

When problems occur in a software project, they almost always require more time, money, or (usually) both.

Sometimes, to solve problems, an additional vendor must be hired or an additional piece of technology must be purchased. If you're using an external software development resource, keeping them on longer will inevitably cost you more.

In the case of an internal team working on software development, the budgetary impact may be masked. You're not paying any *more* for these people to keep working; they're already employed. But, obviously, this is a naïve understanding. Any additional time your employees devote to the software project is time they are not working on something else.

Here's a saying about software development: "The first 80 percent of the project consumes 80 percent of the budget. The final 20 percent of the project consumes the other 80 percent of the budget."

Additional Resources

Often, when a project is beset with problems, a question will come up about adding resources to the project. This happens most frequently when there is a critical date before which the software must be delivered. Business leadership attempts to solve the problem by authorizing more people to work on the project, either by hiring or by asking the vendor to add more people to the team.

Unfortunately, this strategy is not a good one.

There is an old joke in software development about adding additional resources to projects, especially at a late stage, which is when most people try to do it:

First Man: How long does it take for a woman to have a baby?
Second Man: About nine months.
First Man: What if I get two women working on it?

This exchange captures an essential reality of software development regarding timeline. Study after study has shown that throwing additional resources at a software development project, especially in a late stage, does not shorten the timeline in any significant way. And the later resources are added, the less they help.

One reason for this has to do with what I call the "threshold effect." The threshold effect defines a sort of educational chokepoint that all resources must pass through. For a new staff member to be effective on the project, he or she must be brought up to speed on the project. Think of all that information you and your team now know about your project. There are business requirements, visual mockups, modifications, and changes to scope that the original team understands. All that knowledge may be written down—and in a good project it is. But even so, that doesn't mean the information can be instantly absorbed. The existing team has acquired understanding of all this information bit by bit as the project has unfolded. A new resource, or four new resources, must somehow come into possession of all of that knowledge, or a significant portion of it. Who is there to help them? The original team. Thus, the project slows to a crawl as existing team members educate new ones—that is, resources are diverted to deal with the threshold effect.

The second reason adding staff probably won't help shorten the project is what I call the "maximum number of wallpaperers rule." Let's say you're trying to get your house wallpapered, and several rooms have gone well, but there is a big problem in the living room. The project is late, and somehow the wallpaperers just can't seem to solve the difficult angles and non-plumb walls of this particular chamber. Wallpaperers have a lot of equipment: ladders, rolls of wallpaper, paste, long paper-smoothing tools. What's the maximum number of wallpaperers that could fit into the average living room? Five, maybe? Certainly not seventeen. Software development projects, like living rooms, only have so much space. You may have a snarly database problem, but there may only be "room" for one database programmer to work. However long it takes him to get the job done *is* the timeline. Usually there is a kind of ideal-sized team depending on the type of project, just like there is an ideal number of wallpaperers to get that living room done.

Parceling out a project among numerous resources *can* happen. But it must be planned at the *beginning* of development, not at the end. Everyone's part of the dance must be choreographed with everyone else's—like a ballet where the participants' entrances and exits are precisely timed to avoid people crashing into one another. The project itself must be evaluated for the maximum number of resources it can absorb. As we've just seen with the threshold effect, if you are going to have ten people instead of four, they all must be on board at the start, or you will have a big onboarding delay at the end. To address the maximum number of wallpaperers constraint, you would develop a plan of attack for your proverbial living room, with three or four people measuring and checking the odd angles, another two cutting, another one or two hanging, and with the kitchen completely cleared so that all these materials and resources can be deployed effectively.

Trying to figure out how to remain on the original schedule by adding resources at a late stage is a waste of time and energy. This is a fool's errand and will keep you from developing the plans and strategies you need to address your problems.

Fault—A Review

As the old adage goes, "Don't fix the blame, fix the problem." Nothing could be truer in software development. If you're wondering about the cause of your problems, go back and read Chapters 4 through 11. Or, pick from this list:

- Abdication of leadership and decision-making
- Missing, wrong, mixed-up project roles
- Incomplete business requirements; not enough visuals accompanying requirements
- Too many unrecognized risks
- Bad technology team
- Wrong technology choice

Common Late-Stage Problems

There are three common major problem types that emerge at this stage. They are listed here in increasing levels of seriousness:

- Business user revolt
- "Risk chickens" coming home to roost
- "Infections" surfacing

Business User Revolt: "We Talked About It in a Meeting Once"

If you followed the advice in this book to include visuals in your business requirements and provided iterative deliverables and demos, you will have decreased your chance of user revolt somewhat, but not entirely. If you did not, then the business user revolt may be significant. If the project has proceeded along with very little look-see, especially of a visual and clickable kind, the shock and alarm that ensues when the business users get their first look can be extreme. You will hear objections like:

- I didn't think it was going to look like that!
- I thought it was supposed to do such-and-such.
- I never heard that we were going to be doing this-and-that.
- Wait! I don't see a key business requirement. I thought we said in a meeting once that ...

"We talked about it in a meeting once" is one of the most detrimental phrases to software development. Too often, business users expect that if an idea or preference was merely discussed in a meeting, it has become part of the business requirements of the software.

HINT: DON'T SAY IT!

If you are reading this book prior to commencing a software development project, consider mentioning—in a humorous way, of course—that you don't ever want to hear the phrase, "We talked about it in a meeting once." If it's important enough to mention later, it's important enough to write down.

In a full-on user revolt, usually the objections at this stage are exaggerated, as business users declare that the software is "nothing like we discussed" and is "totally unacceptable." On the other side, the technical staff declares that they developed exactly what was defined by the business users.

Managing Business User Revolt

If you have had good, highly visual documentation all along, then your first course of action is to refer your business users back to those documents. Some projects, you may not be surprised to hear, have virtually no documentation or business requirements. It's all been an exchange of e-mails and a bunch of meetings.

Documentation of what the product was supposed to do helps, of course, but it is not a cure-all. As mentioned in the previous chapter, nothing replaces the experience of sitting in front of a computer and clicking. Maybe things just look different on the screen.

Furthermore, certain shifts and compromises will inevitably have occurred in the course of programming. A programmer may have had to make a reasonable judgment call because something could not be implemented exactly as described. This is normal and perfectly appropriate. But, as the programming effort continues, these small shifts and judgment calls build upon one another and can take the business user by surprise. Ideally, business users are kept apprised of any significant changes that occur. But this chapter is about problems, so we can assume that this is not the case.

If even moderate initial documentation exists, the best strategy is to refer business users back to that documentation and walk them through the adjustments or compromises that were not well documented or discussed. Usually, a calm conversation will help stakeholders understand that the team was making its best efforts to balance business requirements, timelines, and the strictures of technology. This will allow everyone to accept what's been done and/or negotiate changes.

It's helpful to be aware that the anger behind business user revolt usually comes from feelings of betrayal or of "not being heard." Referring the user back to the effort that was made in documentation helps to shift those reactions. Many business users have been exposed to the "shaman" phenomenon discussed in Chapter 4. They may be suspicious now that the technical "shamans" are keeping things hidden. They may get frustrated when they are told modifying a feature is not possible because they suspect a "shaman" could do it if he *really* wanted to, but is instead hoarding the magic. This makes it even more important for the project team to be as open and transparent about the development process as possible. Point out baked soufflés and what it will take to un-bake them.

What If No or Little Documentation Exists?

The previous advice only works if you have everything documented in the first place. What if you don't? Well, now it's time for an archeological dig.

The truth is, usually documentation *does* exist, but it may only be found in people's notebooks, e-mails, and memories of meetings. Sometimes, you can document a piece of software after the fact by doing interviews and asking: "What was your understanding of what the software was supposed to do?" Although it is not ideal, after-the-fact documentation can still be useful; often the result is that business users realize the software is closer to their original requirements than they originally thought when they were in that initial state of shock.

In my experience, lack of documentation occurs not because someone was lazy or ignoring his or her job, but because the roles and responsibilities weren't in place. No one realized that documentation was his job in the first place. It's often useful to find examples of good documentation and show it to your team. They may not even know what good documentation is supposed to look like.

HINT: I THOUGHT YOU WERE DOING IT

Everybody always thinks documentation is the other guy's job. It's best to make clear at the start whose job it is.

But let's assume you've reached the problem stage, and it's far too late to teach your team about documentation or to get a business analyst. The best strategy for you now is to collect as much of this information as possible, organize it, and have a frank discussion about what it says. Usually this boils down to:

- Areas where clear common agreement and repeated themes can be identified
- Areas where the documentation is conflicting
- Gray areas

Once the stakeholders realize how little was documented and where documentation is conflicting, there often develops a sense of common responsibility: "Gosh, the technical team was doing the best with what they had." This helps to get beyond the anger that business users often feel when they think nobody listened to them. It can be very helpful in moving towards a list of fixes, changes, and additional development that will be acceptable to the business users.

One silver lining of these sorts of problems is that they may prompt the business leaders to come up with strategies to avoid similar situations in the future.

Risk Chickens Come Home to Roost

In Chapter 8, we covered five of the major technical items that always introduce risk into a software development project. To recap, they are:

1. Integration
2. Data migration
3. Customization
4. Unproven technology/unproven team
5. Too-large project

As discussed in Chapter 8, the best-case scenario is for these risks to be identified up front and for different approaches and Plan Bs to be discussed. As project unfolds, there should be a constant consciousness of these risks. There is a sense of risks always being visible "in the sideview mirrors." Everyone is aware the project team has *consciously* signed up for certain challenges that must be faced. There will be continual checking in, such as: "Hey, where are we on our thinking about that integration?" or, "Has anyone looked at that legacy data yet?" This kind of conversation is a sign of a healthy project.

However, risks are called *risks* for a reason. Even if everyone is aware of the risks ahead of time, they can still cause problems. At the first-look stage of the project, it will become obvious to you which of your "risk chickens" has "come home to roost." Maybe an integration isn't happening as seamlessly as you'd hoped, or perhaps a particular customization can't actually do what was promised. Even in a project where the risks were identified up front, you may face the uncomfortable reality that a business requirement must be compromised, or that getting it done will simply take longer than expected.

Unfortunately, it is often the case that the project team signed up for risks in an unconscious, sometimes heedless, way. As previously mentioned, key words to watch out for are "just" and "no problem." Some examples:

- Integrations are glossed over. ("We'll *just* write a web service that does such and such.")
- Data migration is taken for granted. ("At the end, we'll *just* normalize, de-dupe, and migrate the data.")
- Customizations are minimized. ("That custom widget will be *no problem*.")

Managing the Risk Chickens

If the team realizes risks too late, quick and critical assessments must be performed. Above all, you must find out the answer to the following vital question: "Is solving the problem merely a matter of additional work, or is there a question of actual feasibility?"

In many cases, for example, with tricky feature implementations, the path forward may be more lengthy and require more work than anyone estimated, but there is a *high degree of likelihood* that it will be successful. In other words, you discover your Ikea shipment doesn't contain a desk, but rather a wall-sized entertainment center. It may have many more pieces than you anticipated, and maybe a few are missing or from a discontinued batch, and substitutions must be made, but the likelihood is that the furniture *will* be assembled eventually.

To be clear, more time usually means more money, and this may cause issues with your boss and your boss's boss. But as you make the case, you will have the benefit of knowing that *time* is all you need and the eventual outcome is all but ensured.

It's much worse if feasibility itself is in question. Remember those heart surgeries? Let's say a customization or an integration isn't working. Furthermore, no one is clear on how to *get* it working. There are debates about approach and technologies, and people are hurling shamanistic jargon. This is the most dangerous place to be because you know you will need more time (and more money), but you can't even be sure with more time and more money that you will be successful.

What is required is a *careful* consideration and analysis of the situation. You are looking for an approach that the team feels has the highest probability of being successful. The absolute key here is to reduce or eliminate the uncertainty to the greatest degree possible. It is time to get features and business requirements down to the bare minimum. Feasibility is king. As the saying goes, "It's better to have an okay something than a perfect nothing."

People will say things like, "But if we do it that way, features will have to go out the window!" When you hear this, consider that the survival of your project is at stake. Remember, we don't mind complicated tasks (Ikea desks) or arduous ones (piles of snow). We do mind heart surgeries with a high degree of uncertainty. In Ikea desk and pile of snow situations, more time will be required. But when feasibility itself is still in question, you have no idea how much. Therefore, it is critical to get the answer to the question of feasibility before proceeding.

Seek out the SneakerNet integrations and the Fred Operating System (see Chapter 10 for definitions of these concepts). Negotiate with business users, unhappy though they may be. Feasibility must be assured or nearly so. It is the lifeblood of the project.

This is often the time where business stakeholders get confused and focus on the wrong things. As mentioned before in Chapter 1, people can get very worried when they see the size of the piles of snow or the number of pieces in the Ikea desk. This can distract them from the fact that the programming team doesn't actually *know* how to get the piece of custom code central to the project working. To solve problems, you must focus on getting all parts of the project into the feasible and known category.

If your project is very large but has few other risk factors, you may simply need more time. Team members may not have been able to get their heads around how much there was to do. Maybe they grossly underestimated the amount of effort it would take to get everything done. If your programmers are, say, 40 percent too optimistic (a pretty common phenomenon, as I've noted), then this 40 percent factor tends to compound as more and more tasks are added. Believe it or not, this is a good position to be in, comparatively speaking. Knowing your project is feasible and accomplishable is the most important thing.

However, large projects are often also very complex. They are more likely to involve multiple integrations, custom technology, and data migration. Because of all these pieces, a large complex project is the kind most likely to go off the rails.

This is an opportune moment to reemphasize the benefit of phasing and chunking strategies to make your large project smaller. While some readers may be picking up this book in an emergency situation where problems already exist, others may be in the (fortunate) circumstance of reading this book before attempting a software project. If that is you, listen up!

In Chapter 3, we spoke about phasing and piloting. The idea is to find ways to break your large software project into mini-projects that can be launched one or two at a time. Even though this can produce a disjointed effect, it is much better than the opposite: a large, complex software project beset with problems. I have also mentioned that folks are overoptimistic at the beginning of projects. Remember, optimism is not your friend. You are allowed to hope for the best, but please plan for the worst. Facing a very large project, look for ways to break it down.

However, if you *are* facing a large and complex software project beset with problems, you need to identify areas where feasibility is in question. Those must be resolved first. It may also be helpful to look at strategies for a phased launch. Maybe a section of the software is more ready to go live than others. Consider launching this section. It is an "after the fact" way to make a project smaller, but it can be very effective.

When Programmers Ask for More Time

Often programmers will say they simply want more time to determine if something is feasible. "I just need another week to get it working." This is all right, as long as you set strict boundaries around it.

Check progress in a week. If breakthroughs have been made, allow another week—but meanwhile, Plan Bs should be discussed and alternative approaches identified. Do not allow the situation of "we need more time" to go on indefinitely. It usually takes no more than two or three weeks to get to a point where you can see the light at the end of the tunnel. This means that in two or three weeks you will be able to see that a task is *feasible*, even if it is not finished.

Lurking Infections

In Chapter 8, in addition to risks, we reviewed the "diseases" that can infect projects. These were:

1. Bad technology team
2. Wrong technology choice
3. Lack of leadership

If these infections weren't diagnosed and treated early, your project may have a full-blown flu.

Bad Technology Team

In Chapter 4 I said, "The only vendor that works is a vendor that's worked before."

What's the result of a bad or inexperienced team? Bad code. As noted, it is very hard to determine code quality—even if you have programmers doing code review (that is, one programmer vetting another's code). I have seen projects that did not have a formal code review process go wrong. I have also seen projects *with* a formal review process go sideways. Admittedly, the first is more likely, but the second also happens, usually because a team with poor coders also has poor coders doing review.

How do you even know if you have bad code? It's a good question, and one that is fairly hard to answer. Most coders will find problems in another coder's work. Here's a good rule of thumb: As you emerge into the first-look stage and head towards stability, it should feel like, "Two steps forward, one step back." As one thing gets fixed and done, it may cause another thing to break (the "regression errors"), but once that regression error is fixed, the feature becomes solid.

What it should *not* feel like is, "One step forward, two steps back," an infinite "cha-cha-cha" where stability is never achieved.

If you are constantly feeling like the team is taking one step forward and two steps back, if things are constantly broken and breaking and you can't seem to reach a place of stability, then you can be pretty sure you have a sloppy, incompetent, or simply inexperienced software team.

How to Manage a Bad Technology Team

Here's the bad news: You can't. There is only one thing to do when you find yourself in this situation, and that is to fire the team as quickly as possible. If your internal team is the one with problems, then they must be terminated and a new, more competent team hired.

On the other hand, if you are working with an external vendor, things are a bit stickier. The vendor will do whatever they can to avoid termination, promising fixes and remediations. Often such a vendor has an "A-Team" hidden somewhere in the back room. They propose bringing on the A-Team to rescue the project. But I have encountered problems with this approach. Here's why.

At this point, the vendor is going to be seriously dipping into their profit margin or outright losing money. They have delivered an incomplete and broken product to you. Your schedule is likely blown. Without a doubt, your CFO is going to be having some tough negotiations with them, looking for claw-backs, reductions, and so forth. On top of that, the vendor is now going to have to commit its best, moneymaking A-Team for a significant period of time. Now, how is that *really* going to work?

In the real world, I have seen the A-Team swoop in for a while and make things somewhat better. Then, as the vendor gets tired of making amends and doing penance, you get less and less attention from the A-Team. You may think you are getting a better financial deal—getting the A-Team to fix it all for the same budget—but this rarely actually happens. The vendor will claim it's fixed, or fixed *enough*, and you end up in disputes and sometimes lawsuits.

In my experience, a cleaner and better solution is to negotiate fee concessions from the old vendor and get a new vendor. Unfortunately, fee concessions will be largely dependent on how good your documentation was initially.

Wrong Technology Choice

Making the previous situation worse, it often happens that an incompetent technology team had a hand in the organization's selection of a poor or inappropriate piece of technology.

When the organization finds itself with broken first-look stage software, often fresh outside eyes are brought in. The new tech expert says something like, "Wow. I've never seen SharePoint used to do that!" Or, "Wow, I suppose you could write an accounting system in WordPress, but why would you want to?" A wrong technology choice means that some critical business requirement *cannot* be accomplished with the selected technology or that it would be very difficult to do so.

Managing a Wrong Technology Choice

When a company realizes that they have made a poor choice of technology at the outset, an evaluation must be made of the options going forward. Here are the options with pros and cons:

- Scrap the project and re-approach it with a different technology.
  - *Pro*: Often the project is much more doable with a different technology.
  - *Con*: Budget. You may be lucky and only spend 50 percent of the original budget to redo, because you have already done all the business analysis and documentation. But you will be re-spending for development.
- Keep the technology in place and accept the compromises. Find a way to paper over whatever business requirements cannot be addressed.

- *Pro*: You will bring the project across the finish line, probably close to the original budget.
- *Con*: The business will feel ongoing resentment towards the piece of technology that does not accomplish some key things that were desired.

In my experience, better results are achieved by scrapping the initial technology choice and starting out with a new technology.

The Sunk-Cost Bias

There is significant data showing that people who have invested money in a project have a bias towards investing more money. Your mother or grandmother may have called this "throwing good money after bad." In extreme circumstances this can lead to spending multiples of the original budget. Believe it or not, businesspeople have proven themselves more likely to spend two or three times the original budget *instead of* pulling the emergency brake halfway through, scrapping what's been done, and beginning anew, resulting in a 1.5-times spend.

Avoiding the sunk-cost bias trap is the major reason I prefer the approach of starting anew.

Lack of Leadership

Technology projects require strong, involved leadership. They are not served well by "management by committee." Projects without strong leadership or with big "peanut galleries" often suffer from the arrival of multiple "risk chickens" and/or lots of blossoming infections.

It probably comes as no surprise that inadequate leadership is quite difficult to address. There is no "list of fixes" for the project team. The solution must come from above.

Managing Lack of Leadership

Senior executives are the ones who must decide to replace the project leadership. In the real world, it may not actually look like a pure "leadership issue" because of all the other problems involved. Often the project leader is taken off the project while, at the same time, vendors are fired and project plans are scrapped. To the organization at large, it looks like the project failed for multiple reasons. But, often with a new leader in place, a project can get off to a fresh start with many original barriers swept away.

CHAPTER 13

Launch and Post-Launch: UAT, Security Testing, Performance Testing, Go Live, Rollback Criteria, and Support Mode

People tell me that launching a piece of software, especially a big project, is one of the most stressful things in their professional lives. The planning and work is tremendous. The number of details that must be managed is mind-boggling. When the finish line comes into view, project team members tend to see it as a magic threshold where worries will dissolve. All the uncertainty will be over, and they can finally go out to dinner with their spouses, have real weekends, or go on vacation.

Unfortunately, "launch" never means instant relief. Sometimes there is even a ratcheting up of work and stress. It's best to prepare yourself, your project team, and the business at large for this reality.

User Acceptance Testing: What It Is and When It Happens

In Chapter 11, we talked about quality assurance (QA). To review, this is the process by which an internal or external group (depending on what's right for the project) runs through "happy-path" and "non-happy-path" use cases, finds bugs and fixes them, and then fixes the regression errors. At a certain juncture in the project, another type of testing is needed: user acceptance testing (UAT). Sometimes UAT comes before QA. Other times a bit of QA may be necessary for the product to be ready for UAT. In iterative situations, UAT may have happened bit by bit all along the way.

UAT generally refers to a kind of testing done by the broader company, the "non-techies," who are stakeholders in the project, the customers or customer proxies, whose stamps of approval must be obtained before the software can go live.

HINT: PLAN FOR UAT AHEAD OF TIME

It's best to identify these UAT people at the beginning of the software development process, as they will need to agree to set aside time to review the software and to be trained on methods of testing.

It is great if UAT can happen all along the way in your project, as you present frequent iterative deliverables or demos. As bits of the software get finished, those partially done sections are tested by the business users, and feedback is collected and integrated. This type of UAT can happen even though the software is incomplete and may still be buggy. While it's great if you can work this way, it's not always possible.

For one thing, it may simply be impossible to adopt this method inside your organization. Also, there are times when the only practical way to do UAT is near the end of a project. If the project has many complex pieces and a large data migration aspect, these may all need to converge before a user can reasonably test the software. Such is often the case in a large database or rollout of a piece of financial software. Business users can't "test drive" such software until all the data is in and all the functions are working. Further, what if the business stakeholder group is particularly nontechnical? They might be confused by small bugs and unable to leap over gaps in partially finished software. Therefore, the right time for UAT is largely dependent on the nature of your product and your business users.

As mentioned previously, the UAT people, like your internal QA people, will need to be trained on how to report bugs and other feedback.

Controlling UAT and "We Talked About It in a Meeting Once," Part Deux

Opening up a piece of software to do UAT often causes the technical team to grip the edges of their desks in panic. Because if UAT is not well structured, the process can get out of hand and the software project can grind to a halt with the finish line in sight.

If, for the reasons described earlier or other reasons, your project has not involved a lot of back-and-forth with business users, the software may have been not just out of their sight but also out of mind for a very long time. It may have been months since they looked at the specifications. They have forgotten about the requirements and, dangerously, they may have filled in gaps with their imaginations of what the software *might* do. In Chapter 12, I discussed the kind of business user revolt that can occur when imaginations meet reality.

But even if you don't have a full-on user revolt, you can have some tough times in UAT. The process can trigger tons of feedback, some appropriate and some inappropriate. Stakeholders who have not been involved day-to-day often pop up with "we talked about in a meeting once" comments.

Meanwhile, the project team can also cause problems at UAT time. They are tired and ready to be done. Often they expect the UAT process to be a kind of "rubber stamp," maybe with a tiny tweak here or there.

Both sides must be managed. On the one hand, the business stakeholders doing UAT must review the specifications and agree to boundaries for feedback. You must conduct training sessions with them and create documentation for this purpose.

Classifying UAT Feedback

I have found it is helpful to give UAT testers the following categories to classify their feedback:

1. *Bug*: A feature that breaks or produces an error
2. *Function not working as expected*: A piece of functionality that does something other than what was defined in the business requirements
3. *Request for improvement*: A request to make something work more smoothly with fewer clicks or less confusion
4. *Feature request*: A request for an additional feature to the software

As we'll see, classifying feedback into these four categories is essential for making the conflict resolution process easier. The definitions categorize things in a way that people can easily distinguish "need-to-haves" (1 and 2) from "nice-to-haves" (3 and 4).

The rule of thumb is that all bugs and any functions truly not working as expected get addressed. A small number of requests for improvement are included, while the rest is deferred for post-launch.

HINT: NEED VS. WANT

If the difference between "need-to-have" and "nice-to-have" is introduced early on, testers begin the process in a more helpful frame of mind. You will break the bad habit that many of us have of thinking that every piece of feedback is equally critical.

Bugs

This one's easy. Everyone knows when a button doesn't click or a report calculates the wrong number. While it is the job of the QA team to catch bugs, inevitably UAT testers will run into some, too.

Not Working as Expected—The Trickiest Category

"Feature not working as expected" is the trickiest of the four categories. For that reason, it's important that items falling into this category be clearly differentiated from bugs, improvements, and new features.

As you can imagine, there is often a lot of gray area around "feature not working as expected." For instance: not working as *who* expected? And what exactly *was* expected? Very few business requirements are so thoroughly documented that everyone is crystal clear on the step-by-step behavior of each function.

In "feature not working as expected," one of two things is always true:

A. A business requirement was missed or misunderstood. An intended outcome does not occur.
B. The feature accomplishes the intended outcome, but goes about it in a way that the UAT people find highly objectionable. There may even be an assertion that the method will prove impossible for end users.

Unfortunately, "A," a missed business requirement, sometimes happens even in this late stage of software development. It is why, in Chapter 6, I emphasized that requirements will be re-re-re-understood throughout the life cycle of the project. The software team may have believed in good faith that they were accomplishing the business goal. But maybe some subtlety was missed and no one realized it. It is usually something that seems obvious to the business user—for example, "Everyone knows that the logistics department must have Form A before Form B"— but is not obvious to the technical staff.

Do everything you can to quash statements like, "She should have known," or "I can't believe no one took the time to explain that." Divergences in understanding *do* happen, and it's best to adhere to the adage, "Don't fix blame, fix the problem." It is time to re-understand the requirement and allow programming time to redo the feature. Contingencies in budget and timeline exist, in part, for this reason.

In possibility B, the feature does accomplish its task, but the UAT folks object to the way the feature does it. Examples here may include: The task requires too many screens, or the screens are confusing, or the workflow seems clunky. Here, you will want to ask if the workflow is truly impossible for users to deal with. An example of "impossible" might be if the software is intended to run on a tablet in the field, but requires such large data upload/downloads that the workflow is truly unrealistic over a cell phone connection.

If the function *is* workable, it's time to classify this feedback as a "request for improvement."

Request for Improvement

"Request for improvement" means a function accomplishes its goal in an acceptable way, but a cleaner or easier method can be identified. The business user makes a suggestion for the software to work in the better way. Interestingly, this may be a slap-your-head moment for the project team as they realize, "Wow, that really would work better."

Socialize the idea early on that requests for improvements are nice-to-haves, not need-to-haves. When presented with a request for improvement, the first step is to evaluate the level of effort it will take to complete it. Abide by the following rules of thumb:

- Make improvements if they require straightforward programming time and do not impact the stability of the software. Examples here might include changing wording, drawing attention to a function with a colored button, and putting up an error screen or instructional screen to help a user if there is potential confusion.

- Do not make improvements that require too much "touching" or coding, as they will require additional QA and risk destabilizing the software. If business users are insistent, the matter should be raised to the steering committee.

Feature Request

When the business user identifies a new feature or function that he thinks would make the software better, it's referred to as a "feature request." My advice is that you not tackle feature requests now. Adding features to a nearly done piece of software is a bad practice, because it introduces risk. Who knows what unanticipated consequences on the entire system this new feature will have? Seek high-level approval from the steering committee if business users are adamant (e.g., "Let's raise this to the boss and see if she thinks the feature request should be included.") Usually business users understand the previous points. Often they agree to put a requested feature on a post-launch development list.

HINT: WORKING WITH VENDORS

If you are working with an external vendor, that vendor will help you draw the line with feature requests because doing them will cost money. Usually a vendor will be happy to make small improvements for the good of the product.

Conflict Resolution and Final Launch List

As noted, using the definitions for classifying UAT feedback can make the difference between a conflict-ridden, chaotic UAT process and a smooth one. By putting each finding in the right bucket, the ones that involve the least discussion and conflict resolution can be quickly addressed.

The launch-critical list coming out of UAT should look like this:

- Bug fixes
- Changes to "feature not working as expected" when a business requirement was truly missed or the function workflow is legitimately impossible

- A small number of improvement requests
- No additional features unless authorized at the highest level

Load Testing

Many software projects must consider "load"—that is, the number of users who will be simultaneously using the software when it's launched. This is commonly a factor in launching consumer-facing websites and is handled by the programming team in concert with the systems administrator and often the architect (see Chapter 4 for definitions of these roles). But the truth is, load must be taken into consideration even with internal business software that has small numbers of users. Depending on the software, even a dozen users pounding simultaneously on a button is enough to cause a bottleneck.

Load testing employs tools to simulate the expected load on the software. When slowdowns and chokepoints are discovered, the machines and code can be optimized to relieve them.

Performance Testing

Performance testing generally has a slightly different connotation from load testing because it refers to responsiveness and integrations.

Let's say your project has several integrations—places where one system must talk to another. These two systems may be "talking to one another" across the Internet. There may be a database involved, where a piece of data needs to be looked up after a user makes request. All of these little ask-respond actions, especially across a network or the Internet itself, can cause delays. Code may need to be optimized or more resources brought into play.

Security Testing

Especially in the last few years, security has become a more and more important topic for corporations. News stories reveal breaches in private computer data on a regular basis, and we have no way of knowing how many other breaches occur that are not publicly revealed. It is certain that many companies' computer systems are breached without the companies themselves even knowing it.

The topic of security is a book unto itself. For this book, we have a guest article by cyber security expert and trainer Chris Moschovitis, CSX, CISM.

CYBER SECURITY AND SOFTWARE DEVELOPMENT

Chris Moschovitis, CSX, CISM

You may be amazed to know that it is only recently that cyber security has become an agenda item in software development meetings. As a matter of fact, the majority of popular applications have been developed with little concern about cyber security. Take one of the most popular mobile operating systems: Android. It took several versions before data encryption was even an option. The same is true for Apple's iOS.

Today, following numerous cyber attacks on prominent businesses and institutions, cyber security professionals are starting to have a permanent seat at the table. From the moment the business starts discussing a new software application to the time when the application is running in a stable production environment, cyber security must be involved, and moreover, it must have sign-off authority. Perhaps you think that this is too much. A bit too paranoid? After all, you're only putting up a website, or rolling out an in-house accounting system. Why do you have to be saddled with one more checkpoint, one more expense?

Let's start with some definitions. What *is* cyber security anyway? There are many definitions, but in essence they all boil down to this: Cyber security is the set of best practices and necessary technologies to protect information technology (IT) assets from threats. IT assets encompass all hardware, software, and data either stored or trafficked. Threats are any number of actors intending to damage, steal, or compromise the integrity of assets, even if unintentionally (as in the case of an unsuspecting employee clicking a malware link). In short, it's a dangerous cyber world out there, and your assets are at risk!

As you have already read, managing a software project involves managing risk. Anna has done a brilliant job explaining the risks involved and navigating them. She has taken us on a journey that started at the software project's inception and ended with its successful delivery. My goal in this inset is to take you on a different journey, and this journey starts at the boardroom. Don't worry if you don't have a board. You have something equivalent. Perhaps there are one or more owners. Perhaps you have a managing partner and an executive team. No matter what you call it, there is a set of business executives who have the fiduciary and legal duty to establish what in risk circles is called "residual risk."

Think of it as establishing the cost of doing business, or "acceptable risk." It is a baseline above which the business will not spend money to mitigate risk. Confusing? Consider this: You are walking up to an intersection in New York City. The walk signal is against you. You are alone, and no cars are coming. What do you do? Your decision right there is truly binding! Do you cross? Do you wait for the light to change? Your decision constitutes your acceptance of residual risk. Perhaps you won't see the car rapidly approaching the corner behind you. Perhaps no car is there. Either way, by your decision to cross you accept the outcome. Now, what if you are not alone, but instead you are there with your seven-year-old child. What is your legally binding fiduciary duty then? What would the police say if because of your decision there was an accident and someone got hurt?

Analogous situations occur in business. Here, residual risk is defined as risk minus the application of controls. What is a control? Controls are what reduce risk. In this book we learned of a great control already: Have a super project manager! She is a terrific control in reducing your software development risk. Cyber security professionals know and can deploy a panoply of controls to reduce cyber risk. They can be the right combinations of hardware, software, processes, and training. But—and here is the catch—the right amount of controls is determined by the business. It is the business that decides what acceptable risk is. No one else. It is against this established residual risk framework that cyber security professionals will now work with your software development team to implement controls in each phase of the project, starting with feasibility. Once that's established, cyber security needs to be present and included throughout the life cycle of the project. Let's take a quick look how this would unfold across a typical software development project:

Phase I: Requirements   In this phase a cyber security professional takes a dispassionate view of the project and looks at exposure. Here she will consider data sources, data trafficking, data quality and makeup, and data integrity requirements. She will do so guided by the accepted residual risk framework and decide, for instance, if the data needs to be encrypted "at rest," or if it is adequate for the data to be encrypted "in transit." She will develop a threat assessment based on the proposed application's exposure. For example, is the application open to the Internet? Is there interaction between the application and secure back-end systems? Or, is the application for internal use with no path to the outside world (a.k.a. "air-gapped" application)? Different applications will have different cyber security implications, and it is important to get a good understanding of the exposure up front.

Phase II: Design   During this time, the role of the cyber security professionals is less hands-on, acting more as a member of the steering committee making sure that the cyber security recommendations agreed upon during the requirements phase are being "baked-in" to the design. It is entirely possible that during this phase, depending on the type of project, a cyber security professional may be directly interfacing with systems architects and systems administrators ensuring that the architecture being contemplated satisfies the security requirements of the project.

Phase III: Implementation   Similar to the design phase, the cyber security professionals continue contributing as members of the steering committee, ensuring that the security mandates are being incorporated in the implementation of the system. Of particular concern is the cataloguing of any "back doors" that the programmers may be establishing to assist with their development. Back doors are common and necessary in software development. They allow for very important short cuts and workarounds during implementation and testing but, of course, represent a very serious cyber security vulnerability. Cataloguing all the back doors during implementation will prove invaluable once testing is complete because it is at that point that the programmers will be required to remove them, and the system may have to be audited against them prior to moving to production.

Phase IV: Testing   It is during this phase that the cyber security team earns their money! All applications must undergo thorough vulnerability testing and the overall architecture must go through penetration testing prior to being certified as "production-ready." There are many third-party services and software tools that the cyber team may use (e.g., Burpsuite, Metasploit, etc.) to make sure that the application meets the agreed-upon security threshold. Depending on your industry, third-party certifications may be required at this point. For example, if the application is accepting, processing, or storing credit card data, you may need to ensure that it is payment card industry (PCI) compliant, or in the case of patient information, Health Insurance Portability and Accountability Act (HIPAA) compliant.

Phase V: Maintenance   Once the application has been delivered to production, then the cyber security team integrates it into their "protected suite of assets." It becomes part of standard monitoring and logging, and depending on the environment, shielded through the use of firewalls, penetration detection systems, and potentially incident response systems. Moreover, cyber security needs to be integrated with change management so that any updates or fixes to the application meet the appropriate cyber security thresholds and do not expose the company to any new cyber risk.

Finally, your cyber liability insurance will need to be updated to include the new application that is now in production. This last step is critical and often overlooked. To avoid falling in that trap, make it part of your sign-off checklist. It would be a very rude surprise if after all the effort, time, and expense you suffered a loss that was not covered simply because of an oversight.

For many of you all this cyber security jargon and workflow may sound way too complex and onerous. Perhaps your organization may not even have a formal cyber security team or risk management function. Perhaps you're somewhere in between. That is not what is important. What is important is to recognize that all our business projects must now be examined under a cyber security lens. From the mundane office relocation to exciting and innovative projects like software development, there can be no effort that touches the company that can bypass cyber security examination. Certainly, if the function does not exist in-house, you should consider retaining a cyber security consultant to work with you and your project team to ensure that you have taken all reasonable precautions and steps.

One last thing: Of all the tools, policies and procedures, mandates, and guidelines that both cyber security professionals deploy and organizations need, nothing has proven more effective against cyber threats than awareness training. As a risk control mechanism its effectiveness ranges between 30 percent and 40 percent! If there is only one thing that you can do for your project and for your company, then let it be this: Institute regular cyber security awareness training.

Sign-Off

Here's another *unbreakable* rule: All software launches in an incomplete and buggy state.

It's easy to say, "Of course!" to this statement. Many have read the Leonardo da Vinci quote that art is "never finished, only abandoned." But things sure feel different when it's your own software project. It can be very hard to say "okay" to launch when you know you are still having regression errors, or when a tricky but noncritical bug remains unresolved.

Early on in your project, it's a good idea to identify the group of people who will have the final sign-off on the software. It will be this group's responsibility to decide what level of "bugginess" and incompleteness is acceptable. In general, software is launch-ready when all show-stopping bugs have been removed, UAT feedback has been addressed to the degree possible, all happy paths are clean, and most non-happy paths are reasonably clean.

The acceptable degree of cleanness depends on your company and your risk tolerance. In general, risk-accepting technology start-ups are more likely to launch with software in "beta" stage—a public acknowledgment that it's a work in progress and users should proceed at their own risk. Traditional companies often cannot accept these levels of uncertainty.

Questions to Ask Regarding Launch Readiness

You may find the following questions familiar, as they were included in Chapter 11 with regard to deciding on levels of QA:

- Will bugs in the software be embarrassing, damaging to your corporate image, or harmful to your relationship with your customers?
- Is this a financial data system where loss of data is unacceptable?
- Is the software critical to company operations?

If the answer to any of these questions is "yes," then QA must be longer and bug tolerance lower.

Not Knowing Is Not Acceptable

Way back in Chapter 4, I spoke about the need for business stakeholders to embrace their role as project decision makers. I encouraged these folks to get support, ask questions, and unmask "shamans." Indeed, empowering nontechnical businesspeople is one of the main reasons I am writing this book.

Right now, the involvement and understanding of the key business stakeholders is critical. One of the most common things I have seen in software development projects is for the lead business stakeholder to be blind to the risks at launch. He or she is not capable of adequately answering the list of questions presented earlier. Understanding the risks to data, financial systems, corporate image, and business processes requires in-depth discussions about databases, integrations, and other factors. Gaining such understanding often involves hearing a programmer say, "We cannot reproduce the behavior documented in Jira ticket AB204." Businesspeople, who can be very tired by the end of a project, may not have the patience to dig into what this means. If the business stakeholder has shown impatience and intolerance for technical conversations, the programming team may have learned that certain topics are simply off limits. The result? The tech team will actively conceal information.

How does this sad tale end? With the business stakeholder relying 100 percent on the programming team's word for when launch is or is not appropriate. Here's the problem with that scenario: If the software is launched and a bug compromises some critical business function, the business leader's only answer to her boss, the board, and the shareholders is that she simply did not know. Don't let this be you!

It takes patience, but with some determination you will be able to gauge the level of risk and select an appropriate time to launch.

Criteria for Rollback

It is sometimes the case that a software project goes live and must be rolled back. "Rollback" refers to the process of switching to the previous software system because some critical flaw exists in the new software system and/or its infrastructure. It is a last-resort strategy when a system is failing.

Even in healthy software projects, the need for rollback can occur. It is not common for businesses to roll back due to a light or even moderate number of bugs.

The following rollback criteria are the most common:

- Load and performance are seriously compromised. The new system is so choked with traffic or poor performance that customers cannot use the system.
- Critical business functions are blocked. Examples of this might be reports with important due dates that cannot be delivered to customers.
- Financial operations are affected.

Each business and software project may have its own additional rollback criteria. The project team should discuss ahead of time what will constitute a problem serious enough for rollback. The rollback criteria should be approved by the steering committee.

After a rollback, the project team will develop a plan to remediate the issues, and a new go-live date will be set.

Singing the Post-Launch Blues

The period immediately following launch can be one of the most chaotic and stressful times in your project for the following six reasons. It is wise, as part of training and internal communication, to inform business stakeholders of these six points well ahead of time, so that they are prepared:

1. *Bugs*: No QA team can ever replace dozens, hundreds, or thousands of actual users. They will find bugs you didn't know you had.
2. *Creative use*: "I can't believe the users are doing *that!*" is a statement often heard post-launch, as users find ways to operate the software that were never intended or imagined by the project team. Often this "creative use" can cause breakages, data corruptions, malfunctions, and user-support issues.

3. *Load/performance*: Load, or the number of users the software has to accommodate at any given time, is one of the hardest things to simulate accurately. Ditto with performance over a network. For one thing, if you are launching a piece of software online to the general public, you may have no idea how many users are going to come. But even if your predictions are accurate in terms of the number of users, you may not be able to estimate *how* people are going to use the software. If users are running live reports and doing lots of data uploads and downloads, your newly launched software may face traffic and performance demands it's not prepared for.

4. *Negative impact of change*: Change is hard for most human beings, even if it's a change for the better. Many software projects have positive change as a key goal, such as re-launching a website or re-doing an accounting system, and the reasons for the change may be obvious. But when the new software comes out, you may have a lot of users express longing for the old systems and processes, even if they were broken, time consuming, or out of date. It's a "devil you know" kind of thing.

   For publicly facing software, like websites, site traffic may actually decline. Users may not have liked the old site, but they knew how to get around it! For internal business software, it may take users longer to do their daily tasks until they become comfortable on the new system.

5. *Customer support*: Customer support usually increases right after a software launch. And yet, businesses often forget to even tell customer support that software is being launched and is likely to impact their lives. Don't be that person: Give your customer support team a heads-up.

6. *Operating in a live state*: In pre-launch, it's uncomfortable to discover bugs. But when the site is live, it's downright painful. In a live state, a bug is much more critical, especially if it's blocking users from doing what they want. The ratcheting-up of urgency once you go live can be enormously stressful.

HINT: LEARNING AND RE-LEARNING

Training for internal users often happens too early on to be helpful. If a month or two has gone by since training, users will forget what they learned. Consider offering booster-training sessions immediately prior to launch.

Was It All a Big Mistake?

With the disruption invariably caused by a project launch, the discovery of new bugs, issues with load, and negative feedback from people who hate change, it's no wonder that this is a moment when some folks will question whether the whole project was a big mistake. In fact, your boss or internal stakeholder group may quickly demand an after action review (AAR) or some other kind of post-mortem analysis of the project to determine what happened.

It certainly takes more than a couple of weeks after launch to decide on the success of a project. If possible, delay the analysis until things have had a chance to settle down. Maybe those users who are struggling with the early stages of learning a new piece of software will feel much more comfortable after a month. Website traffic patterns also take time to normalize. Avoid reactive decision making until the "burn in" period is done.

Metrics

In Chapter 6, I recommended establishing metrics for your software project's success as part of your business requirements. These might include things like improved user satisfaction, increase in efficiency (e.g., less time is needed to do something), more data, more accurate data, and positive impacts on business drivers such as website traffic and sales. Now is the time to start collecting information on the things that you decided to measure. The process of collecting data can help calm folks down if you are getting an "it was all a big mistake" reaction.

Sometimes it is necessary to do a kind of "post-mortem" review because your boss requires it. In my view, these are often done too soon, when emotions are still running high and there is not enough distance and perspective. It's often useful to agree to feedback, surveys, and other forms of feedback, even in this too-soon time frame, as long as you insist in further feedback at a later date—say, six months from project launch.

Ongoing Development

In the months immediately following launch, project teams often face as much work as they did in the lead-up to launch. There is the post-launch "punch list," which includes still-open but noncritical bugs, as well as bugs found after launch, plus requests for improvements and features that cropped up in UAT.

Now is the time to develop a process for managing ongoing development. The Agile method, mentioned in the Introduction, has great strategies for managing ongoing development: A list of all improvements and bugs is made and quickly estimated by the programming team. Many teams choose to use very broad strokes in estimating, such as one hour, half day, full day, half week, and full week.

As tasks are estimated, the tech team and the business users *together* prioritize what is to get done first. The incorporation of business users into the prioritization of development work often produces a very efficient and collaborative environment where everyone knows what's important to get done and why.

Surviving the Next One

This book has provided a framework for managing software projects—the things I have seen time and time again in website launches, database migrations, finance system replacements, custom application development, and packaged software deployments. But your project was undoubtedly unique. What did you learn? What new adages could you coin?

As with scenes in dreams, the events and lessons of your project will be easy to forget. Take an hour to write them down. Another useful step—put out a survey specific to your team and collect their observations, impressions, and lessons learned.

You'll thank yourself when the next project must be done.

Project Tools

1. Project Roles—Checklist and Blend-ability
2. Budgeting Formulas—Calculating a Budget Estimate
3. Budgeting for Contingency—Arriving at a Contingency Number
4. Project Meetings—Key Meetings, Participants, and Agendas
5. Running Effective Meetings—Tips for Keep Meetings on Track
6. Trough of FUD—Graphic of Emotions in Software Development
7. Alpha Stage/First Look—How to Distinguish a Good Mess from a Bad Mess
8. Project Timeline—A High Level Typical Timeline
9. Heat Map—A Tool to Track Project Status
10. Budget Tracking—A Tool to Report on Project Budget
11. Project Flow Graphic—A Graphic Showing Times of Project Conflict and Calm
12. Common Late-Stage Problems—The Three Most Common Causes of Problems
13. Classifying UAT Feedback—Instructions to User Acceptance Testers
14. Cyber Security—Important Safety Tips

1. Project Roles—Checklist and Blend-ability

The following worksheet outlines the roles necessary on a project and which roles may best be blended:

Role	Blend with?	Notes
Program manager	Business analyst	CIO, CTO, or consultant
Project manager	Do not blend	Detailed thinker. Choose PM with good software development background as opposed to industry-specific.
Business analyst	UX and program management	BAs are big-picture thinkers. A visually trained BA can do UX. Sometimes a vendor's account manager will also serve as a BA.
User experience	BA or front-end programmer	A visual BA can also serve in the UX role. Also, a front-end programmer with good communication skills can play the UX role.
Designer	UX	A designer who is very experienced digitally can sometimes make the transition to UX.
Front-end programmer	UX	If the front-end programmer has good communication skills, he or she can sometimes do UX tasks.
Back-end programmer	Architect	A good back-end programmer may be able to recommend systems architecture.
Architect	Back-end Programming	A good back-end programmer can play this role.
Systems administrator	Architect	A good systems administrator can also play the architect role.

2. Budgeting Formulas—Calculating a Budget Estimate

The following three principles can be used in creating a budget:

1. The software development budget should reflect the following percentages on the four prime components of development.
   - Programming features: 60 percent
   - Additional discovery: 15 percent
   - Unit testing and debugging: 10 percent
   - Project management: 15 percent
2. First, obtain initial programmer estimates, then adjust them.
   Take the initial programmer estimate as the "raw" hours. Increase that number by 40 percent, or some known error factor, to get the final estimate of hours.
3. Create the budget by calculating the total hours estimate, then work backwards.

Explanation of the math:

In the case where 70 hours is the final adjusted programming estimate, those 70 hours represent 60 percent of the total project.

To derive the total project estimate, use the following mathematical statement:

Seventy hours represents 60% of *what number?*
Put algebraically

$$70 = .6 * X$$

Solving for X

$$70 \div .6 = 116.66$$

Round up to 117 hours.

Now, the other numbers may be calculated as percentages of the total, 15 percent for additional discovery, 10 percent for debugging, and 15 percent for project management.

The following spreadsheet uses the previous budgeting formula and is available at http://emediaweb.com/completesoftwarepm.

Project Budget	Programmer 1 Backend	Programmer 2 Front End	Total Raw Hours (+40%)	Total Size if Programming is 60%	Discovery	Unit Testing	Project Management	Programming ($150/hr)	Discovery ($135/hr)	Unit Testing ($150/hr)	Project Management ($125/hr)	Totals
Feature 1	7	0	9.8	16	2.45	1.63	2.45	$ 1,470.00	$ 330.75	$ 245.00	$ 306.25	$ 2,352.00
Feature 2	2	4	8.4	14	2.1	1.40	2.1	$ 1,260.00	$ 283.50	$ 210.00	$ 262.50	$ 2,016.00
Feature 3	0	8	11.2	19	2.8	1.87	2.8	$ 1,680.00	$ 378.00	$ 280.00	$ 350.00	$ 2,688.00
TOTAL												$ 7,056.00

3. Budgeting for Contingency—Arriving at a Contingency Number

The following contingency percentage factors may be used to estimate a project's contingency budget:

Type of Risk	Suggested Contingency
Tricky data migration	+10 percent
Integration	+5 percent each
Large, "un-chunk-able" project	+15 percent
Unproven technology	+25–50 percent
Customization/innovation	Assigned in proportion to the amount of customization/innovation involved (e.g., a project with 20% innovative features means a +20 percent contingency)

An example based on a large project with two significant integration points:

Initial Budget	$250,000
plus large project factor	*$37,500*
plus integration factor	*$25,000*
plus innovation factor	*$12,500*
Total contingency	*$75,000*
Total budget	$325,000

4. Project Meetings—Key Meetings, Participants, and Agendas

The following are key meetings recommended through the life cycle of a project.

Project Kickoff Meeting

The project kickoff occurs when the project scope has been defined and development tasks are about to commence. It "kicks off" the project in a public and official fashion business-wide.

Participants

- Project leadership

  Project sponsor: The key executive in the business who has responsibility for the project

  Program manager/account manager: The overall lead on the project; uber-project manager

  Project manager: The person in charge of the project breakdown, all the day-to-day tracking tools including ticketing systems, wikis, and budget reporting tools

  Team leads: Often includes the lead programmer, main interface designer, and vendor representatives

  Internal stakeholders: The internal customers for whom the project is being delivered

- Company leadership (as needed)
  - CEO
  - COO
  - CFO
  - CIO

Agenda

- High-level project definition and project charter
- Business case and metrics
- Project approach and technologies
- Introduction of team members and roles
- Project scope
- Out of scope
- Timeline
- Budget and budget reporting
- Risks, cautions, and disclaimers

Monthly Steering Committee Meeting

This is a monthly meeting of the project governance team. The project steering committee makes decisions about the project affecting features, budget, and timeline. It has the following recommended participants:

Participants

- Project sponsor
- Key business stakeholder
- Program manager
- Project manager
- Tech lead

Agenda

- Work completed—previous month
- Work to be completed—coming month
- Budget status
- Timeline status
- Hot topics affecting important features, timeline, or budget

Weekly Project Management Meeting

A comprehensive update of the project, track by track.

Participants

- Project sponsor (as needed)
- Program manager
- Project manager
- Vendor project managers
- Programming leads

Agenda

- Reporting on all tracks of the project with discussion and problem solving as appropriate
- Identifying to-dos and follow-ups
- Planning tasks in the upcoming week

Daily Standup Meeting

A daily work-status meeting.

Participants

- Project manager
- Tech team
- Other attendees affecting work plan as necessary, such as designer, sys admin, or business analyst

Agenda

- Status report from each member: tasks accomplished, plan for the day
- Report of barriers and concerns

5. Running Effective Meetings—Tips to Keep Meetings on Track

Here are some recommended ways to achieve well-run meetings.

Insist on Attention

Set boundaries around multitasking. State up front that the meetings will be well run and purposeful. Tell your team members that you'll keep your end of the bargain if they keep theirs—full attention.

Timeliness

Do not wait for late participants. Start the meeting on time to encourage members to show up on time.

Avoid Too Much Getting "into the Weeds"

Avoid discussing the "how." Update meetings are not about finding solutions, but about identifying problems and high-level follow-ups that lead to a solution.

If in-the-weeds conversations crop up, listen to see if there is some critical point that needs to be addressed. If not, recommend a follow-up meeting with a tight group to discuss.

Identify When Things Need to Be Kicked Upstairs

Project teams should not spend time talking about strategic business issues that are not in their purview. Be able to identify those topics that require a higher level business decision. Make plans to recommend an approach and follow up at the appropriate business level.

Fix Poor Quality Sound—Speaker Phones and Cell Phones

Make decisions that allow for the best sound quality possible. Speaker phones should be of high quality or not used. Cell phone connections should be good, and you should be in an area free of noise.

Avoid Too Much Talk, Especially by the Leader

Meeting leaders should deliver a paragraph of information, then pause and ask for direct feedback. If it is necessary to deliver a lot of information at once, alert the audience that the leader will engage in a longer "frame-up" or "context setting."

Provide Agenda and Take Notes

All meetings should have a published agenda and follow-up notes. Have one person regularly tasked with taking the notes and distributing them. A good set of notes records the key points made under each agenda item. The notes end with the follow-up steps and who is on point to address them.

6. The Trough of FUD—Graphic of Emotions in Software Development

Originated by the Gartner Group, the Hype Cycle graphs the elevated and then deflated expectations that happen with regard to technology initiatives. (For more detail, see Gartner's explanation of a hype cycle at http://www.gartner.com/technology/research/methodologies/hype-cycle .jsp.)

A software-development graph illustrates the Trough of FUD. "FUD" stands for fear, uncertainty, and doubt. This related tool tracks the emotion of a software development project against the objective reality of the software.

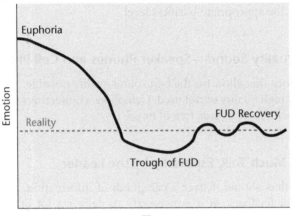

7. Alpha Stage/First Look—How to Distinguish a Good Mess from a Bad Mess

The first look at a software product (in the alpha stage) can be messy. Be concerned about messiness if:

- What's messy and broken is the simple stuff.
- The workflows aren't in place or don't make sense.

Other Hints:

- The tech staff should not be allowed to save all the difficult features for later.
- It should feel as if the technical staff is grappling with the harder challenges such as discussing different approaches and testing them out.
- An exception is data migration, which almost always must occur toward the end of the project.

8. Project Timeline—A High-Level Typical Timeline

The following is a high-level timeline of how projects usually unfold:

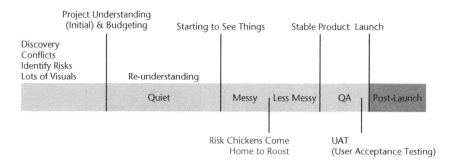

9. Heat Map—A Tool to Track Project Status

A heat map is a combination timeline and project status tool. All the tracks are laid out with delivery dates shown by the end of a colored bar. In the future, the bar is gray. The bar gets filled in as weeks proceed with colors.

If a project track is going well, the present week will be filled in with green. If the track is at risk, yellow is used. If a date has been missed or a budget overrun has occurred, the track is filled in with red.

As new dates or budgets are agreed upon, the track returns to green.

A fully completed heat map (below) shows the history of the project. A sample heat map is available at http://emediaweb.com/completesoftwarepm.

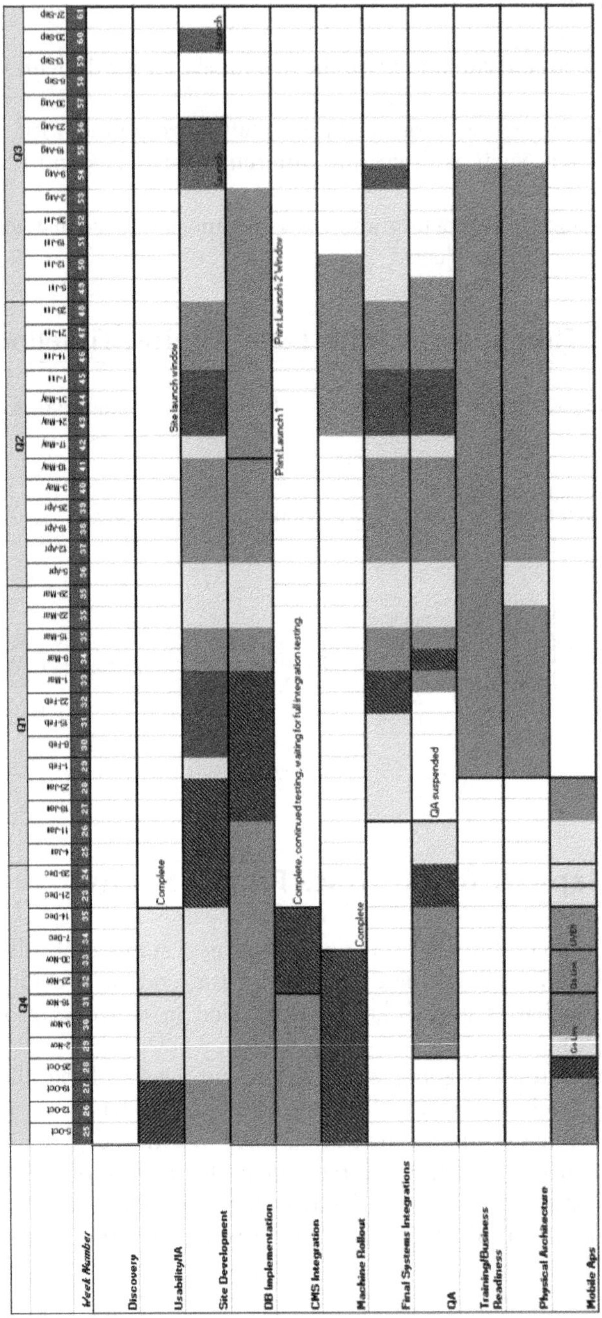
Q4
Q1
Q2
Q3
Week Number
Discovery
Usability/IA
Complete
Site Development
Site launch window
DB Implementation
CMS Integration
Complete, continued testing, waiting for full integration testing.
Pilot Launch 1
Pilot Launch 2 Window
Machine Rollout
Complete
Final Systems Integrations
QA
QA suspended
Training/Business Readiness
Physical Architecture
Mobile Apps

10. Budget Tracking—A Tool to Report on Project Budget Status

Week by week, the project budget is tracked. The following reporting tool may be used to track budgeted expenses against actual expenditures. A sample tracking tool is available at http://emediaweb.com/completesoftwarepm.

Item	Budget	Actual	Variance
Planning & Documentation	$5,500	$5,000	($500)
Programming	$25,000	$33,000	$8,000
Data Migration	$7,500	$7,000	($500)
Design	$4,000	$4,000	$0
Hosting Setup	$3,500	$2,500	($1,000)
Licenses	$6,000	$6,000	$0
TOTAL	$51,500	$57,500	$6,000

11. Project Flow Graphic—A Graphic Showing Times of Project Conflict and Calm

This image shows that projects start with many conflicts and decisions. Afterwards the project unfolds pretty quietly. Towards the end of the project risks surface, all the trains come together on the tracks, and there can be problems.

Project Beginning   Project Middle   Project End

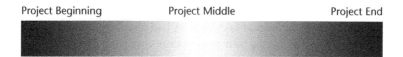

12. Common Late-Stage Problems—The Three Most Common Causes of Problems

There are three common major problem types that emerge at a late stage of a project. They are listed here in increasing levels of seriousness:

- *Business user revolt:* Business users reject the software saying it doesn't meet specifications.
- *"Risk chickens" coming home to roost:* The consequences of risks, known or unknown, emerge in a project.

- *"Infections" surface:* Bad teams, strategic disagreements, or other lurking unresolved conflicts bubble to the surface and compromise the project.

13. Classifying UAT Feedback—Instructions to User Acceptance Testers

The following categories are helpful for UAT to classify their feedback:

1. *Bug:* A feature that breaks or produces an error
2. *Function not working as expected:* A piece of functionality that does something other than what was defined in the business requirements
3. *Request for improvement:* A request to make something work more smoothly, with fewer clicks or less confusion
4. *Feature request:* A request for an additional feature to the software

14. Cyber Security—Important Safety Tips

1. Make sure that cyber security is present and signs off throughout the project's life cycle, starting with feasibility, requirements, and design all the way to proof of concept, development, testing, deployment, and maintenance.
2. Don't forget: Old applications don't die. They become cyber security vulnerabilities. Manage end-of-life systems carefully.
3. If the application is mission-critical, engage in third-party code review.
4. Change management is critical. A bad change management process will not only break your production environment, but may expose you to cyber security vulnerabilities. This is especially applicable to the inevitable "emergency change management" all too common in application maintenance.
5. Beware software utility tools! Keep these under lock-and-key. They are like construction tools left lying around. The same hammer that built your project can be used to destroy it.
6. Make sure you perform a risk assessment and a business impact analysis. They will help define threats, recovery time objectives (how long you can afford to be down), and recovery point objectives (how much data you can afford to lose if you must roll back).
7. Perform vulnerability scanning on all platforms: development, staging (testing/QA/UAT), and production.

8. Make a practice to continually check with the Open Web Application Security Project (OWASP) at https://www.owasp.org, and Carnegie Mellon's CERT division of software engineering at http://www.cert .org for the most recently discovered software vulnerabilities.
9. Perform penetration testing to expose unremediated vulnerabilities.
10. Never forget that managing cyber security risk is an ongoing process. You're never done. Integrate it in all your processes and remain diligent.

GLOSSARY

Agile methodology: A method of developing software with limited up-front definition and budgeting; the Agile process values constant collaboration, frequent deliverables, and continuous evolution of requirements.

Alpha: The first view of a piece of software; software is able to be reviewed and initial testing begun.

Architect: The person responsible for the choice, organization, setup, and configuration of the hardware that will run the software

Back-end programmer: The person who programs databases and system integrations

Beta: The second phase of software delivery; all features are complete and the software is ready for comprehensive testing and debugging.

Blended budgeting: The process of using two or more budgeting methods to estimate a project

Blue sky: Brainstorming about possible solutions and strategies

Bottom-up budgeting: The process of using a detailed task list of which the project is comprised and estimating costs based on that list; the total budget is a sum of individual costs.

Budgeting: The process of estimating costs for a project

Business analyst: The business analyst (BA) is the person who collects, digests, and codifies all the software development requirements from the business stakeholders.

Business case: A high-level statement of the project's purpose, including return on investment (ROI); the business case identifies how the project will benefit the business, often through potential profits, cost savings, and/or competitive advantages gained through the completion of the project.

Business requirements: A list of features and functions required in the software project

Business stakeholder(s): The people inside the business, often referred to as internal customers, for whom the project is being delivered

CIO: Chief information officer, usually the most senior executive in a business responsible for technology; a strategic role

Comparative budgeting: The process of using a previously completed project to estimate costs for a new project

Comparison grid: A method of comparing software solutions or vendors; a list of qualities to which scores may be given

Complex: A term used to describe certain project activities; complex activities are difficult to understand and hard to execute and require a high degree of specialized expertise. Unknown variables and conditions that may affect the outcome are present.

Complicated: A term used to describe certain project activities; complicated endeavors have many steps, require a detailed set of instructions to implement the steps, and involve specialized expertise. Misunderstandings may occur. Unknown variables are not present.

Contingency: An amount reserved in the project budget to address risks, both known and unknown

CTO: Chief technology officer, a senior executive within an organization responsible for technology. In comparison with the CIO, the CTO's role is usually more operational and/or executional. The CTO often reports to the CIO.

Current state: The existing state of software systems. Such systems may also be called legacy systems.

Customization: Writing software code that is original or unique; also, making changes to existing code (such as to a software package or open-source module) so that it becomes original or unique

Custom software: Understood in contrast to off-the-shelf software; custom software is coded from scratch or by using a software development framework.

Daily standup meeting: A meeting of everyone involved in the project execution phase to track daily progress, decide on work plans, and identify barriers to work

Data migration: The transfer of data from one system into another system

Debugging: The process of correcting flaws in a software product

Designer: The team member who creates the visual look of the software

Discovery: A research phase of a project where possible solutions and approaches are examined

Documentation: Any and all documents relating to the project

Framework: A software development platform that provides greater speed in creating custom-developed software; common code, frequently available in an open-source community, provides generic functionality, feature modules, and other instruments that can be used in the creation of new software products.

Fred operating system: A term specific to this book; engaging a human being to substitute for a software feature

Front-end programmer: The team member who programs the interface layer of the software

Happy path: A route through the software that is the most logical and common; using the software as intended

Hardware: The machines, physical or virtual, on which software runs

Heart surgery: A metaphor used in this book to illustrate the nature of complex tasks

Heat map: A method of tracking the main project tracks; a heat map typically uses green to indicate a track is on time and on budget, yellow to indicate it is at risk, and red to indicate it has gone over schedule or over budget. A heatmap example is available at my website http://www.emediaweb.com/completesoftwarepm

Hosting: The service associated with keeping a website or other software system running on a server, physical or virtual

Infrastructure: The set of hardware and software components, physical or virtual, that enable an application to run. Exists within a hosting environment.

Integration: The exchange of data between software systems

Ikea desks: A metaphor used in this book to illustrate the nature of complicated tasks

Insource: The use of internal staff for roles on a project

Key business stakeholder: The person inside a business, often referred to as the main internal customer, for whom the project is being delivered

Latency: Delays across networks such as local area networks or the Internet

Licensing fees: The fees that must be paid for the use of third-party systems, products, or software

Load testing: The process of testing how much traffic the software can simultaneously support

Metrics: Often called KPIs, or key performance indicators; metrics are variables that can be identified and in which a change can be measured. The set of variables, taken together, will suggest whether or not the project has achieved the goals laid out in the business case.

Monthly steering committee meeting: A meeting of the project governance team, the steering committee, to review recent work, upcoming work, and any project risks, barriers, or shifts. It is the responsibility of the steering committee to make decisions affecting feature set, timeline, and budget.

Moore's Law: The axiom stating that computing power doubles every 18 months

Off-the-shelf software: Understood in contrast to custom coding; pre-built software that is installed and configured with little original or from-scratch work

One-time setup fees: Costs, incurred only once, to set up a new system, product, or server infrastructure

Outsource: To obtain external resources for roles on a project; the term "contract" is often used.

Performance testing: Related to load testing; performance testing evaluates the responsiveness of a piece of software, typically including the speed of the interface, media loads and data loads.

Phased approach: The approach by which a piece of software is delivered in planned releases or phases

Piles of snow: A metaphor used in this book to illustrate the nature of simple tasks

Pilot: The release of a software project in which one or both of the following is true: (1) The software is released to a limited group of users; (2) The software has a reduced feature set compared to the final specification.

Product roadmap: A plan for future software development, often by the makers of off-the-shelf software; the product roadmap sets out the timeline for the release of features, patches, and updates.

Program manager: Usually seen on medium-to-large projects; oversees the entire project team and coordinates with the business

Project approach: The method of tackling a software project; different approaches include phased launches, piloting, custom development, and use of off-the-shelf software.

Project charter: A high-level document that defines the project giving the business case, project goals, and budget

Project kickoff meeting: A meeting held when discovery is complete and project scope is known. It usually includes business stakeholders and project leadership. The agenda includes project definition, metrics, roles, scope, out of scope, risks, budget, and timeline.

Project leadership: The top roles in the project team; usually includes the project sponsor, key business stakeholder, program manager, project manager, and technical lead

Project management: The entire process of managing a software project, soup to nuts, involving the project team as well as the business at large; also, in a more limited sense, the discipline practiced by a project manager

Project manager: A professional in the field of project management in charge of planning and tracking the execution of software projects; often possesses certificates such as PMP

Project scope: The set of features that will be delivered in the project; also called business requirements

Project sponsor: The business manager charged with overseeing the software development project and with the accountability for its success or failure

Project tracks: A breakdown of the project into logical segments

Quality assurance testing: Formal testing of software involving the writing and running of test scripts and the documentation of bugs or flaws

Regression errors: Bugs or flaws that are introduced into a piece of software as a consequence of fixing other bugs or adding features

Regression testing: Software testing to find regression errors

RFP: Request for proposal, a formal document produced by a company seeking estimates on software; the document contains all the features and functions desired. All vendors are considered to be on an equal playing field. Questions are asked and answered in controlled ways to make sure no one vendor is in possession of more information than another.

Risks: In software, known and unknown issues that are likely to cause time and budget overruns in the project such as complex integrations, data migration, and customizations

Scope creep: Gradual small changes in project scope that, taken together, can add up to a large project impact in time and budget.

Scope document: Also referred to as the business requirements document; a detailed exposition of the software's features and functions that will be used as the definitive point of reference by team members to execute the project

Security testing: Testing a piece of software for vulnerabilities that can be exploited by malicious actors

Sign-off: The official process by which the end customer accepts a piece of software

Simple: A term used to describe some project activities; simple endeavors are easy to conceptualize and straightforward to perform. They do not require a high degree of specialized expertise, just "elbow grease."

SneakerNet: A manual integration, such as e-mailing a set of data or uploading it

Software development: Any activity that involves the installation, creation, or customization of software such as launching a website, installing a new accounting system, or building a custom application

Software development life cycle: Represents software development like the circle of life, including analysis, design, implementation, testing, and evolution

Stakeholder: A person in the business at large who is an internal customer for the software project

Steering committee: A project governance body that includes project sponsor, key business stakeholder, and project leadership; it makes significant decisions about a project such as changes to timeline, budget, and major feature additions or subtractions.

Subject matter expert: A person with vertical expertise in a certain area

Sunk-cost bias: The tendency to invest more time and money in a project that is not going well instead of canceling the project

Support: Addressing user requests and fixing flaws in a piece of software that is live

Systems administrator: Responsible for the ongoing upkeep, maintenance, and reliable operation of the machines that run the software product

Timeline: The schedule for the project; the project timeline often has many mini-timelines for the major tracks in the project.

Top-down budgeting: Usually used in large projects where it is impractical or impossible to arrive at a detailed list of tasks and where comparison budgeting is not possible; a holistic budget arrived at through estimating large chunks of a project

Unit testing: Testing done by programmers to ensure a feature works

Upgrade path: The planned upgrades in a software system, software framework, or operating system; upgrades may introduce new features, take advantage of new hardware developments, and patch security vulnerabilities. They are often phrased as version releases such as 1.0 and 2.0.

User acceptance testing: A form of testing done by a small group of customers of the software or proxies for those customers; the user acceptance testing (UAT) group typically tests a beta stage product to ensure it meets the business requirements.

User experience expert: The user experience (UX) person is a professional who designs the navigation and user interaction (i.e., which clicks lead to what) in a piece of software.

Waterfall: A traditional method of software development; projects are completed using a sequential series of steps including initial gathering of requirements, developing of software according to those requirements, testing, and delivering. Changes and shifts in requirements are discouraged.

Weekly project management meeting: A weekly meeting to review progress on all major project tracks, to review status, identify issues that need further discussion, and to give a view into upcoming work.

INDEX

Printed and bound by CPI Group (UK) Ltd, Croydon, CR0 4YY

04/06/2023

03224099-0001